Agents and Actions Supplements

Proceedings of the
Symposium on

Aspirin and Related Drugs: Their Actions and Uses

Editors
K.D. Rainsford
K. Brune
M.W. Whitehouse

Birkhäuser Verlag
Basel und Stuttgart

Edited by:
K.D. Rainsford, Department of Biochemistry, University of Tasmania
Medical School, Hobart, Tasmania, Australia.

K. Brune, Department of Pharmacology,
Biozentrum der Universität,
Basel, Switzerland.

M.W. Whitehouse, Department of Experimental Pathology,
John Curtin School for Medical Research,
Australian National University, Canberra, Australia.

This Symposium was held in conjuction with the Physiology Section of the 47th Congress of the Australian and New Zealand Association for the Advancement of Science on Friday 14th May 1976 at the University of Tasmania, Hobart, Tasmania, Australia.

QV95
.S94
1976

CIP-Kurztitelaufnahme der Deutschen Bibliothek

Symposium on Aspirin and Related Drugs, Their Actions and Uses ‹1976, Hobart, Tasmania›
Proceedings of the Symposium on Aspirin and Related Drugs, Their Actions and Uses/ed. K.D. Rainsford ... This symposium was held in conjunction with the Physiology Section of the 47. Congress of the Australian and New Zealand Assoc. for the Advancement of Science on Friday 14. May 1976 at the Univ. of Tasmania, Hobart, Tasmania, Australia. - 1. Aufl. - Basel, Stuttgart: Birkhäuser, 1977. -
(Agents and actions supplements; 1)
ISBN 3-7643-0902-4

NE: Rainsford, Kay D. [Hrsg.]; University of Tasmania ‹Hobart, Tasmania›; Aspirin and related drugs, their actions and uses

All Rights Reserved. No part of this publication may be reproduced, stored in a retrieval system, or transmitted, in any form or by any means, electronic, mechanical, photocopying, recording or otherwise, without the prior permission of the Copyright owner.

© Birkhäuser Verlag Basel, 1977

PREFACE

For same years there has been much concern expressed by all sections of western communities about the side effects of therapeutic use and abuse of aspirin and related drugs. This is especially so in Australia where it appears that we have a particularly high incidence of gastric ulceration and renal failure (probably amongst the highest in the world) and this has been associated with the high rate of consumption of aspirin-containing preparations. On the other hand; after years of research effort to find replacements for aspirin in the treatment of rheumatic conditions it is clear that aspirin is still the main drug of choice. It was for these reasons that it was felt timely to have a symposium, of international character, here in Australia where we have much dedicated research effort in this field.

The papers feature recent studies on the therapeutic effectiveness, side-effects and biological actions to these effects of aspirin, comparisons have been made with other non-steroid anti-inflammatory/analgesic drugs; an approach which enables an up to date assessment of the best drug for the treatment of inflammatory conditions.

If nothing else, some guidelines for the safer use of aspirin and related drugs by extending their therapeutic rational may have come out of this conference. This may justify publication of the proceedings.

The conference would not have been possible but for the efforts of those who travelled long distances to participate and provided manuscripts within short time for publication. Our thanks to the publishers for arranging rapid publication with attendant benefits. Also thanks to Miss Vreni Forster at the Department of Pharmacology, Biocenter of the University of Basel for valuable help in preparing manuscripts.

Finally we hope that the success of this conference will be extended to ensure that a regular conference will be held every year or two to present our appraisal and future look in research on aspirin and related drugs.

K.D. Rainsford

CONTENTS

Chairman's Opening Remarks — 7
    K.D.Muirden

A Pharmacokinetic Approach to the Understanding of Therapeutic Effects and Side Effects of Salicylates — 9
    K.Brune, P.Graff and K.D.Rainsford

Are there Antirheumatic Drugs which could Increase Prostaglandin Synthesis? — 27
    J.-P. Famaey, J.Fontaine and J.Reuse

Actions of Anti-Inflammatory Drugs on Smooth Muscle — 33
    D.M.Temple and I.A.McKnight

Salicylates in Rheumatoid Arthritis: Pharmacokinetics and Analgesic Response. — 35
    G.Graham

Alternatives to Aspirin, Derived from Biological Sources. — 43
    M.W.Whitehouse, K.D.Rainsford, I.G.Young, N.G.Ardlie and K.Brune

Gastrointestinal and other Side-Effects from the Use of Aspirin and Related Drugs; Biochemical Studies on the Mechanisms of Gastrotoxicity. — 59
    K.D.Rainsford

Gastric Ulcer, Aspirin Esterase and Aspirin. — 71
    K.D.Landecker

Aspirin and Ulcers — 81
    J.Duggan

Interactions of Anti-Inflammatory Drugs. — 85
    P.M.Brooks

Salicylate Toxicology — 97
    E.G.McQueen

Salicylates and Copper: Ternary Copper (II) Complexes Containing Salicylate and Nitrogenous Chelates such as Histamine. — 109
    W.R.Walker and R.Reeves

CHAIRMAN'S OPENING REMARKS

K.D.Muirden
Department of Medicine, University of Melbourne, Royal Melbourne Hospital, Melbourne, Victoria, Australia

Aspirin has a time honoured place in treating rheumatic disorders and in a recent survey of 88 patients with rheumatoid arthritis attending my clinic, 85 % were receiving aspirin. The drug is inexpensive and relatively safe yet its mode of action, pharmacological effects, its method of administration and its toxicity provoke continuous comment if not in fact controversy. In treating rheumatoid arthritis, we believe that to produce optimal anti-inflammatory effects an attempt should be made to achieve adequate serum levels of salicylate. Despite this I discovered this week that in the last 14 patients in whom therapeutic doses were being used, 2 patients had potentially toxic blood levels, in 6 the level was too low and in only the other 6 were we within the accepted therapeutic range. Difficulties were compounded by the use of no less than 6 different forms of salicylate in these 14 cases (and the daily dose ranged from 2.6 to 5.2 gms of drug).

I hope we shall consider today how valid it is to use plasma drug concentrations as therapeutic guides. Are single values of any use? Should we concentrate on early morning levels when the rheumatoid arthritis patient is at his worst and when adequate blood levels seem so necessary? Should our laboratories measure both free and protein bound levels in drugs which are so strongly bound, especially when we are dealing with diseases where serum albumin levels may be low and where displacement due to the use of additional drugs with similar properties is possible? It has been alledged that with indomethacin and with gold salts, serum levels of drug do not relate well with therapeutic and toxic effects. Is this true and can we conceive of an explanation?

When rheumatoid arthritis is not controlled with aspirin, additional anti-inflammatory drugs will be added to the patients' regime. In our 88 cases, only 12 were on one anti-inflammatory drug and the others were on 2, 3, 4 and 5 such medications. To add to the potential for drug interaction, 22 were on additional drugs for unrelated medical conditions excluding antacids, iron and vitamins.

Despite nearly 100 years of use, major discoveries relating to the pharmacology of aspirin continue to be made. We shall be asked today to look at the effect of drugs on prostaglandins and not just as inhibitors of prostaglandin synthesis but also as stimulators. The effect of aspirin on platelet function has forced cardiologists, neurologists, haematologists and even possibly renal physicians to come to terms with the drug. To rheumatologists who have been rebuked for this apparent over and mis-use of aspirin, it seems ironic that its potential for preventing vascular catastrophies may at last relieve them of some guilt.

The relationship between rheumatologists and gastorenterologists remains at best an uneasy one. Most of this afternoon's programme concentrates on this critical interaction. The multitude of different forms of aspirin are partly related to an attempt to overcome this problem. Each new anti-inflammatory drug is advertised as being "better tolerated" than aspirin - meaning for the G.I. tract. Rarely would the advertisement suggest that the new product is therapeutically more effective.

The renal side effects of analgesics have received much publicity in this part of the world and it is appropriate to mention at this ANZAAS conference that important work relating renal effects and analgesics in rheumatoid arthritis involved a cooperative study between centers in Australia and New Zealand. A recent College of Physicians Symposium devoted considerable time to this problem and I appreciate that the organizers of this Symposium today have not felt obliged to over-emphasize these aspects.

We shall, however, look forward to hearing of drugs which are alternatives to aspirin - a veritable flood of new drugs. Drugs which perhaps "spare" corticosteroids; drugs which are effective but claim the compulsory lower frequency of side effects. If we look at my figures again - 50 % of our rheumatoid patients were on indomethacin - a drug which was heralded when first introduced with considerable suspicion, and ibuprofen in 25 % - introduced with meagre enthusiasm by Australian rheumatologists. Both drugs are now well established following the revelation that the initially recommended dose of indomethacin was too high and that for ibuprofen too low. Ibuprofen has now spawned a host of other propionic acid derivatives - naproxen, fenoprofen, ketoprofen - all poised to be released onto this lucrative market.

In discussing anti-inflammatory drugs today could I remind you that inflammation itself may not necessarily be detrimental to the patient and that relief of inflammation itself without knowledge of its cause could well leave us with some misgivings. The auto-immuno phenomena which seem so obvious in rheumatic diseases gives us a theoretical licence to suppress inflammation with some determination, but the concept of immune-stimulation advanced recently should make us pause and reflect once more. We do appear to have two drugs which act in a more profound way on the rheumatoid process by comparison with aspirin and even with the corticosteroids. I refer to the gold salts and penicillamine. I would hope that one of our distinguished speakers today will have time to enlarge on this intriguing and I believe crucial property of these drugs.

A Pharmacokinetic Approach to the Understanding of Therapeutic Effects and Side Effects of Salicylates

by
    K. Brune*, P. Graf* and K.D. Rainsford**

Introduction

It is now almost hundred years since Aspirin was first introduced as an anti-inflammatory agent. Since then the question of how aspirin-like drugs exert their anti-inflammatory action has always been an enigmatic one for physicians and pharmacologists. In the past it was hoped that an answer to this question would come from the discovery of some unique pharmacodynamic action or even a specific "receptor" ([1]) for these non-steroid (so called aspirin-like) anti-inflammatory drugs (NSAID). However, even today after decades of research in this field, a convincing pharmacodynamic explanation is not evident. Serious arguments can, as may be seen from Table I, be raised against each one of the mechanisms of action which have been proposed during the last twenty years. Even the currently popular concept that NSAID act by inhibiting the prostaglandin (PG) synthesis ([2]) leaves some crucial questions unexplained; for example:

1) Why are only acidic compounds useful anti-inflammatory agents in the clinic, although many non-acidic compounds have been found potent inhibitors of PG-synthesis in vitro (see Table I)?

2) Why do only acidic inhibitors of PG-synthesis cause side effects in the stomach and the kidney although both side effects have been claimed to be due to inhibition of PG-synthesis ([3], [4])?

3) Why is the degree of binding to plasma proteins important for the anti-inflammatory potency of NSAID ([5]) when most other drugs are more effective when they show a lesser degree of protein binding?

These unanswered questions prompted us to consider the possibility that NSAID show a specific pharmacokinetic behaviour in vivo which together with some well known pharmacodynamic effects could solve the old question how and why NSAID are useful anti-inflammatory compounds.

---

* Department of Pharmacology, Biozentrum der Universität Basel, Klingelbergstrasse 70, CH-4056 Basel, Switzerland

** Department of Biochemistry, University of Tasmania, G.P.O. Box 252c, Hobart, Tasmania, Australia 7001

TABLE I: CONCEPTS TO EXPLAIN THE MODE OF ACTION OF NSAID AND ARGUMENT AGAINST THEM

Mechanisms implicated in mediation of anti-inflammatory effects of NSAID	References	Argument against these postulated mechanisms	References
Uncoupling of oxidative phosphorylation	12	Some potent uncouplers do not act as anti-inflammatory agents	13
Inhibition of histamine liberation and blockade of tissue response to histamine	14	Histamine is only of moderate importance in inflammation and antihistamines are hardly effective in most types of inflammation	14 15
Inhibition of serotonin liberation and blockade of tissue response to serotonin	14	Only a few inflammatory reactions are mediated in part by serotonin	16
Antagonization of kinin generation and activity	14	Effect is incomplete, irregular and unspecific	14
Displacement of anti-inflammatory peptides from albumins	17	Clearly anti-inflammatory peptides are unknown at present	18 2
Inhibition of lysosomal enzyme release	19	Inverse correlation between lysosome stabilization and anti-inflammatory potency	2 20 21
Hyperpolarization of of neuronal membranes	22	Could only be observed in non-mammalian species	23
Inhibition of prostaglandin synthesis	2 24	Some potent inhibitors of prostaglandin-synthesis *in vitro* are poor anti-inflammatory drugs *in vivo*	25 26
Inhibition of complement activation	27	Importance *in vivo* is uncertain, high concentrations are necessary	27

The Pharmacokinetic Concept

A) Principles

Most drugs show unequal distribution throughout the body of man and experimental animals. Some drugs accumulate in certain areas and show effects in others. These drugs often exert their action by interacting with specific pharmacological receptors while they distribute throughout the body according to some physicochemical characteristics. For these drugs it is only important to reach sufficient concentrations in areas endowed with their receptors. Since these receptors are often only to be found in some areas but lack in others one cannot conclude that these drugs have effects in areas where they reach high concentrations. On the other hand, some drugs do not act through specific pharmacological receptors. These drugs may, e.g. interfere in an unspecific way with ubiquitous cell structures as e.g., phospholipids as in biological membranes. These drugs by contrast will have prominent effects only in compartments where they reach high concentrations.

As pointed out before, we believe there is good evidence to assume that NSAID are such drugs which do not act through one specific, unequally distributed type of receptor. Instead they may unspecifically interfere with membrane functions like some other widely used drugs, which are weak organic acids, namely the barbiturates. They exert unspecific, concentration dependent actions (6). If this assumption is correct one should expect high concentrations of NSAID in compartments where they have effects or, in turn, to have effects where they reach high concentrations.

B) Consequences

If one accepts this assumption then one may look for particular chemical properties of all NSAID which could account for unequal biodistribution leading to localized action. Such characteristics do exist. Firstly, all of these drugs show a comparatively high degree of binding to plasma proteins _in vivo_ (7). Moreover, the degree of binding _in vivo_ correlates well with the efficacy of these drugs in counteracting inflammatory reactions (5). This is an unique finding in pharmacology because normally a high degree of binding to plasma proteins reduces the efficacy _in vivo_! One explanation for this unique feature of NSAID may be that these drugs reach high concentrations in the plasma and leave the intravascular space following capillary damage and leakage of plasma into inflamed sites. Secondly, another physico-chemical characteristic of NSAID namely their common pKa-values (approximate pKa=4) (8,9) will, according to the principles of general pharmacology, lead in part to specific biodistribution of NSAID. According to the concept of non-ionic diffusion, weak organic acids as NSAID will, in compartments separated by biological membranes, reach highest concentration in the most alkaline one. This is normally the plasma and the extracellular space (see Fig.1)(10). However, there are few exceptions. In the stomach, especially the proton secreting parietal cells are directly surrounded

DISTRIBUTION OF PHENYLBUTAZONE

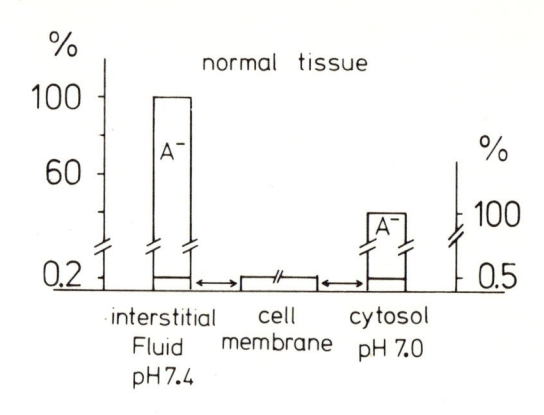

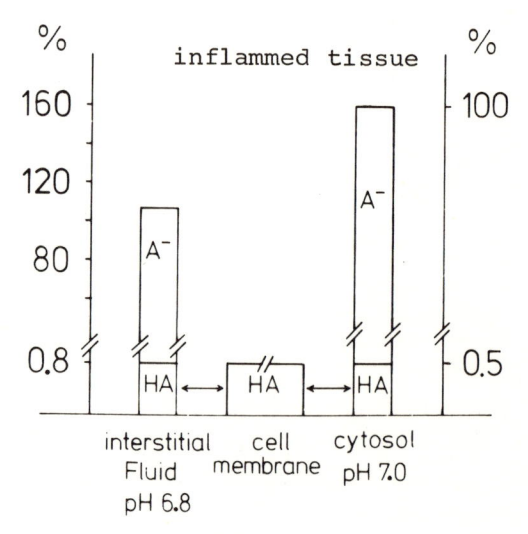

Figure 1: Influence of variation of extracellular pH on the concentrations of phenylbutazone. It is assumed that the cell membrane is impermeable for phenylbutazone ions. Under this condition, the concentration in the intracellular space is a function of the concentration of non-ionized phenylbutazone in the interstitial fluid and the pH in the intracellular space. The figure shows that ion trapping in the cells will occur when the extracellular pH decreases.

by the highly acidic gastric fluid, and they are not separated by a protective mucus layer as other cells (see Fig.2). In addition, after oral administration of NSAID there will be high concentrations of drug in the stomach. In the kidney, the cells of the distal nephrons are in contact with urine which is normally acidic. In man and most animals NSAID will be eli-

GASTRIC PIT   FUNDIC GLAND

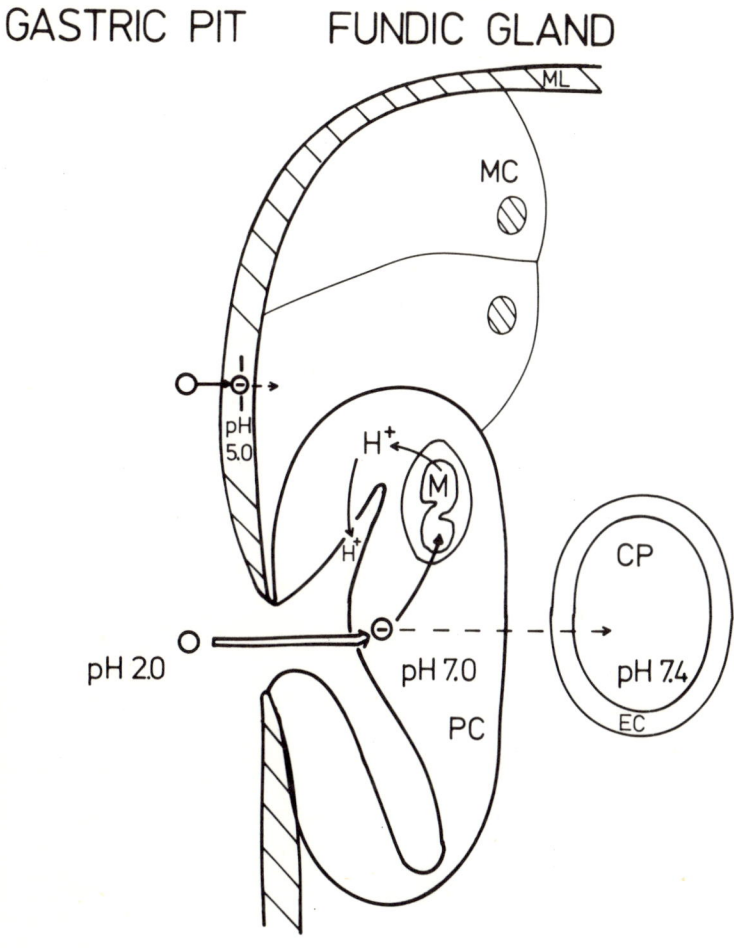

a) In the acidic gastric contents acetylsalicylic acid is non-ionized (O). In this form it enters the mucus layer (ML, pH 5.0) becomes ionized (⊖) interacts with the mucus (φ) and can penetrate less rapidly into the mucosa (-----)

b) Parietal cells (PC) secrete protons (H$^+$) from their canals. From these canals (at acid pH) non-ionized acetylsalicylic acid (O) rapidly penetrates into the intracellular space of PC. There it is ionized (⊖) and so trapped.

Figure 2:
Rapid invasion of non-ionized acetylsalicylic acid (O) in parietal cells of the stomach, uncoupling of oxidative phosphorylation in abundant mitochondria (M) and slow passage of ionized acetylsalicylic acid (⊖) into neighbouring compartments (e.g. endothelial cells (EC) of capillaries (CP) or mucus cells (MC) accounts for gastric damage originating in parietal cells. These cells open a gate into the otherwise mucus protected mucosa.

minated via the kidney and thus eventually reach high concentrations in the kidney tubules. Finally, in inflammed tissue lactacidosis causes a lowering of the pH in the extracellular space (11), so that following capillary leakage high concentrations of NSAID are to be found at these sites. In all these instances drugs will move through the cell membranes into the cell interior, i.e. to a site where they can cause pharmacological actions.

Evidence Favouring the Pharmacokinetic Concept.

To test this concept we performed studies to learn more about the biodistribution of NSAID in inflammation using radioautography and biochemical analysis. Firstly, when ^{14}C-labelled phenylbutazone, indomethacin or antipyrine was administered to young rats, and simultaneously an inflammation elicited by the injection of carrageenan into the left hind-paw and the subcutaneous tissue of the neck, radio-autographs as given in Fig. 3 were obtained. With the acidic NSAID they show high radioactivity in the inflamed tissue of the neck and the left hind-paw, and also in stomach wall the lumen of the small intestine and the kidney. In contrast, experiments using ^{14}C-labelled antipyrine (which is not an acid) do not show accumulation of activity in the inflamed tissue (see Fig. 3c), indicating (as suggested) that only acidic NSAID accumulate in inflamed tissue.

To obtain quantitative information on this process we measured the concentration of different acidic NSAID and their non-acidic congeners in the synovial fluid obtained from inflamed and non-inflamed joints of chickens at different times after drug administration. The results for two indole derivatives and two pyrazolones are given in Table II. Clearly, there is accumulation of the two acidic NSAID in inflamed joints. The concentration of these drugs in the fluid of these joints was about three times higher than in the control joints at three hours, and exceeded the concentrations measured in the blood plasma at this same time. On the other hand, the non-acidic congeners did not show such behaviour. The concentration of these drugs was equal or only slightly higher in the fluid of the inflamed as compared to the control joints and did not reach as high concentrations as in the plasma. Correspondingly, five to ten times higher doses of the non-acidic drugs were necessary to measurably inhibit PG-synthesis _in vivo_ despite almost equal effectiveness of both derivatives _in vitro_ (25). These results show the importance of pharmacokinetics for the antiinflammatory action of the acidic NSAID.

We also tried to get evidence supporting our pharmacokinetic concept of the mode of action of NSAID by investigating the common side-effects of these drugs in the stomach. We came to the conclusion that the proton-secreting parietal cells not only provide the acidic environment and thus the driving force for rapid (non-ionic) absorption of weak acids in the stomach, moreover we assumed that these cells, being unprotected by a mucus layer should serve as an entrance for these acids (see Fig.2). On the basis of these considerations, one should expect high concentrations of NSAID to occur in parietal cells, followed by morphological signs of cell damage shortly after oral administration.

Indeed, it can be shown (Fig.4) that relatively high concentrations of radioactivity are obtained already one minute after oral administration of ^{14}C-labelled aspirin in the glandular part of the rat stomach. By contrast, the non-glandular mucosa of the stomach shows much lower concentrations of the drug. It should be kept in mind that only the glandular mucosa contains

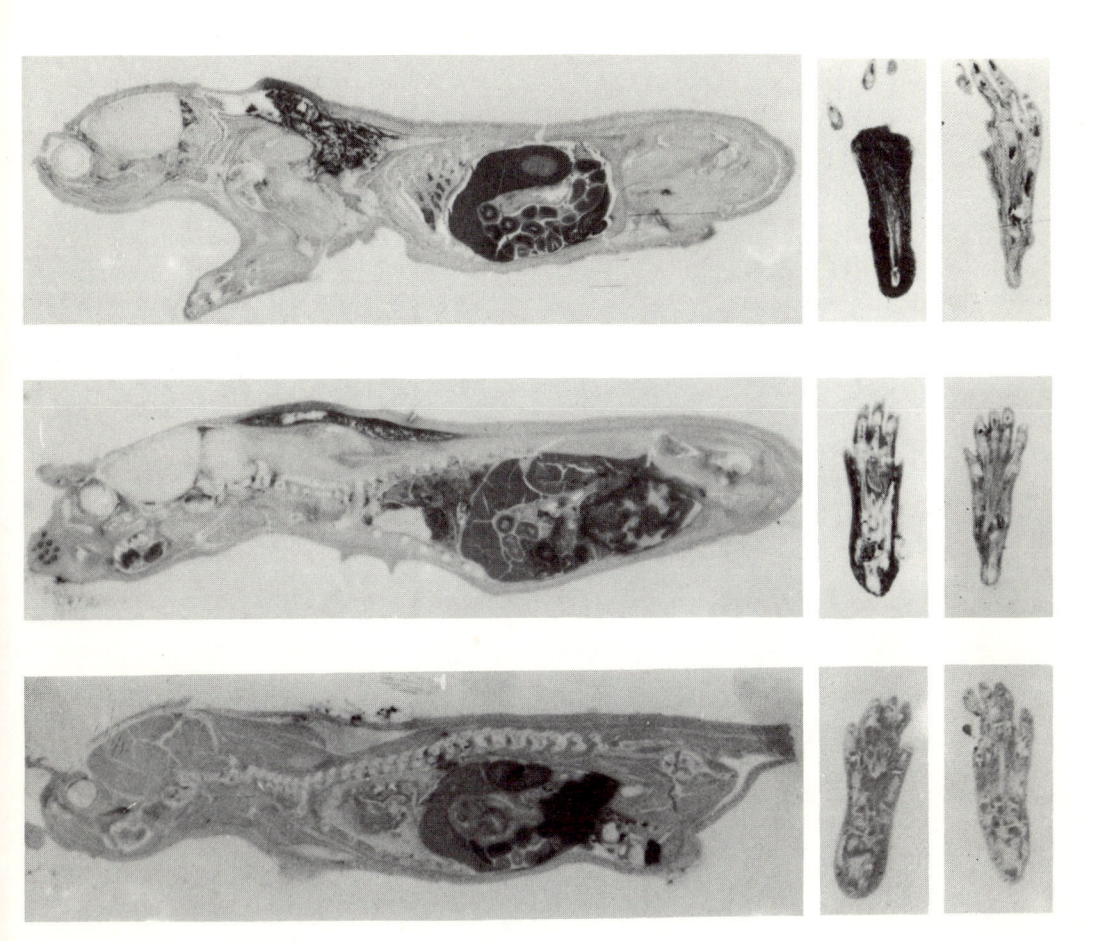

Figure 3: Radioautographs of ^{14}C-phenylbutazone (a), ^{14}C-indomethacin (b) and ^{14}C-antipyrine (c) treated rats.
Inflammation was elicited by the injection of carageenan into the left hind-paw and the subcutaneous tissue of the neck. The phenylbutazone or indomethacin treated animals (a and b) show high activity in the inflammed tissue of the neck and the left hind-paw whilst the antipyrine treated (c) does not display high activity in the inflammed tissue.

parietal cells and only this tissue shows ulcerations after administration of NSAID. At later times, the activity in the non-glandular mucosa reaches higher levels than in the glandular part but never attains the concentrations observed in the glandular part at one minute after administration. This observation corresponds with morphological observations given in Fig. 5 and 6. In Fig. 5 one has the impression that the focal erosion developing one hour after aspirin-administration in the glandular mucosa progresses beneath the surface in the area where many parietal cells are located. Even more convincing are the results

TABLE II: ACIDIC AND NON-ACIDIC PYRAZOLONE AND INDOLE DERIVATIVES: INHIBITION OF PG-SYNTHESIS AND DRUG CONTENT IN JOINT FLUID AND PLASMA 2 HOURS AFTER ELICITING AN URATE ARTHRITIS

DRUG		ACIDITY	DOSE i.v. (mg/kg)	PG F_{2a} - CONTENT (% of CONTROLS)	AMOUNT OF DRUG (ng) FOUND IN: PLASMA (0.3ml)	JOINT WASH CONTROL	JOINT WASH INFLAMMED
PYRAZOLONE DERIVATIVES	ANTIPYRINE	ALKALINE pK_a 4.4	200	51±26.1 *	27.3±8.8	14.8±1.1	18.0±5.0
	PHENYLBUTAZONE	ACIDIC pK_a 4.4	20	39±13.6 *	2.6±0.9	0.7±0.2	3.0±1.1
INDOLE DERIVATES	INDOXOLE	ALKALINE pK_a 2.0	50	23±13.3 *	0.41±0.06	0.16±0.03	0.19±0.02
	INDOMETHACIN	ACIDIC pK_a 4.2	5	25±22.0 *	0.55±0.36	0.20±0.1	0.72±0.33 *

The drugs were dissolved in DMSO and infused slowly (10 minutes) i.v. at zero time. One hour later urate crystals (UC) were injected (4% w/v in saline) into the right intertarsal joint of the chicken (2 kg body weight) the left joint receiving 0.3 ml saline as a control. Three hours later joint washes were performed, the PG F_{2a} content measured in the UC injected joints and the drug dependent inhibition of PG-synthesis expressed in percent of DMSO treated controls. On other animals having received the same treatment the drug content in the inflammed and control joints was measured three hours after drug administration. The values for 0.3 ml plasma are given because a joint wash contained about 0.3 to 0.4 ml fluid, i.e. not much more than was injected two hours before. Means and standard deviations of 5 or more experiments are given,
* p<0.01. For technical details see lit. 21, 28, 29.

TABLE III: EVIDENCE IMPLICATING PARIETAL CELLS IN THE DEVELOPMENT OF GASTRIC DAMAGE BY SALICYLATES

	Clinical Observations	References		Animal Studies	References
A)	Achlorhydric patients develop much less gastric side	31	A)	Damage or inhibition of parietal cells by x-ray, atropine or vagotomy result in reduction of gastric damage.	35
B)	Addition of buffering agents markedly reduces gastric damage.	32	B)	Buffering or chronic administration of salicylates decreases or abolishes gastric damage.	36 37
C)	Factors stimulating production of acid (e.g. alcohol, caffeine, nicotine etc.) increase the incidence of gastric damage due to salicylates	33	C)	Activation of parietal cells by cholinergic drugs increase the incidence of gastric damage	38
			D)	Morphological changes appear only in parietal cells containing areas of the rat stomach	30 34
			E)	Parietal cells show morphological decay already a few minutes after p.o. aspirin administration	30 34

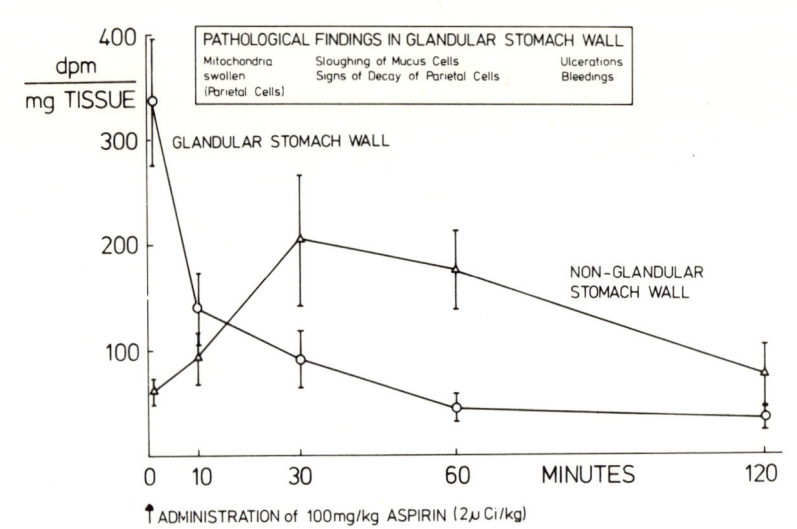

Figure 4: Concentration of ^{14}C from ^{14}C-labelled aspirin in the mucosa of the glandular and non-glandular part of the rat stomach. Fasted rats were given C-labelled aspirin at 0-time. After different time intervals, the animals were killed. The stomach was removed within 30 sec., rinsed for 10 sec. with saline, cleaned with tissue, weighed, digested and samples were counted. The pathological findings were obtained from parallel studies. Highest concentration of activity was observed 1 minute after administration in the glandular part of the stomach.

of electronmicroscopic investigations. Specimens obtained about one hour after aspirin-administration show uniform signs of decay in the appearance of parietal cells in the rat (Fig.6). The mitochondria are swollen and the canaliculi are distorted but these signs only occur in parietal cells whilst other cell types in the same area appear undamaged. Similar results were obtained in the ferret (30). Additional evidence for the importance of parietal cells for the gastric side effects of NSAID such as aspirin comes from clinical and experimental work which has already been published (see Table III).

Finally it shall be pointed out that:

a) There is increasing evidence that "micropharmacokinetical" events leading to accumulation of NSAID in certain cells in the kidney may at least in part be responsible for damage occuring in the tubuli and papillae of the kidney. Some observations are summarized in table IV.

b) It is well accepted that the greatly enhanced toxicity of aspirin in acidotic patients is due to higher concentrations achieved by this drug in the cells of the CNS under acidotic conditions (52). The same holds true for barbitu-

TABLE IV: EVIDENCE IMPLICATING ACIDIC NSAID IN THE
DEVELOPMENT OF KIDNEY DAMAGE

Clinical Observations	References	Animal Studies	References
a) Aspirin causes (transient) loss of tubule epithelia.	39	a) Papillary necrosis in rats, rabbits and mice could be elicited with several acidic NSAID alone	46 47 48 49 50 51
b) Phenylbutazone and other NSAID were shown to cause papillary necrosis	40 41	b) Attempts to obtain kidney damage with phenacetin alone were unsuccessful or led to controversial results	42 43 44 45
c) Most cases of "Phenacetin-nephropathia" were caused by mixtures of phenacetin with aspirin or other (acidic) NSAID	42 43 44 45	c) Combinations of phenacetin with aspirin gave rise to kidney-damage but the results indicate that aspirin is the more important component	42 43 44 45

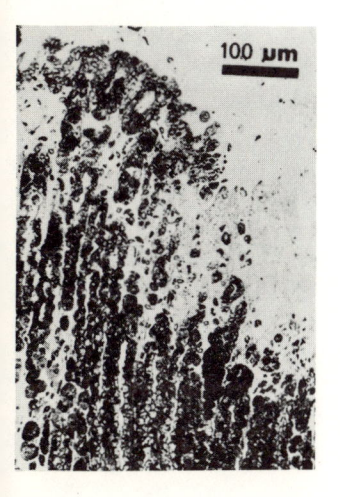

Figure 5:
Erosion in the glandular gastric mucosa of a rat after a single oral dose of 200 mg/kg body weight aspirin. The eroded area progresses beneath the surface. Extensive damage occurs in the area. The erosion stains poorly. The cells deeper in the mucosa below and adjacent to erosions appeared intact and stain normally. X 190

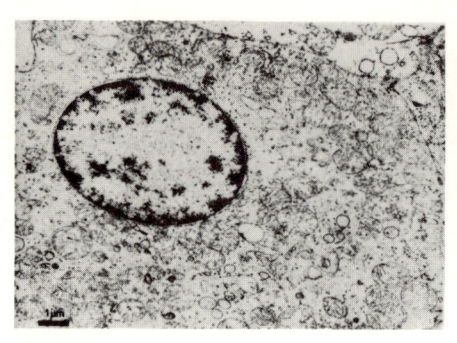

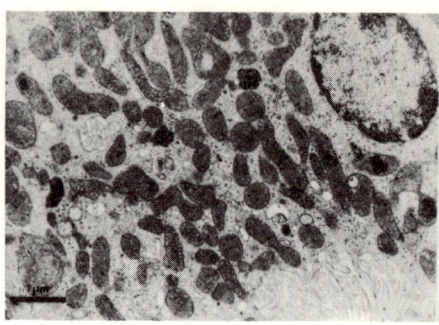

Figure 6

a) Parietal cells in the otherwise undamaged areas of the gastric mucosa of a rat given an oral dose of 200 mg/kg body weight aspirin for one hour showing marked damage to the mitochondria and clumping of nuclear material in these cells. x 16 500.
b) Parietal cell from a control rat given 1 ml H_2O for one hour.

    rats which are shifted into the cell interior under acidotic conditions and reappear in high concentrations in the plasma during alkalosis (53).

c) The comparatively high efficacy of phenylbutazone in man can also be explained by it's unique pharmacokinetics in human beings; phenylbutazone shows a different pharmacokinetic behaviour in humans as compared to most other species. In humans, phenylbutazone is almost completely bound to plasma proteins and its volume of distribution is close to the plasma volume. In addition, it is very slowly metabolized and/or excreted (54). These characteristics lead to long lasting high concentrations of drug in the plasma, making the plasma a reservoir from which, if our concept is correct, drugs will leak into inflamed tissue whereever and whenever capillary damage caused by inflammation occurs. Also, high concentrations of phenylbutazone maintained in the plasma might contribute to the rare but dangerous side effects as bone marrow and liver cell damage since the cells in these organs are blood bathed and not well protected by capillary walls. Most other NSAID show comparatively short half lives in humans which in turn normally leads to very low blood levels between two administrations possibly allowing for recovery of these organs.

Some Unsolved Problems.

There is no doubt that our pharmacokinetic approach to the understanding of the action of NSAID is as yet not completely proven by experimental evidence. Especially the final proof that NSAID accumulate within specific cells in the stomach, the kidney and the inflammed tissue is still lacking. It may come from radioautographic studies on the microscopical level. However, it is quite difficult to get realistic pictures of the distribution of these drugs in microscopical sections because any fixation procedure besides freezing will cause redistribution of substances which are both, soluble in water and organic solvents.

Also the question remains what is (are) the pharmacodynamic events on the cellular lever which lead(s) to e.g., reduction of pain from inflammation and/or cell damage in the stomach? Is it the well known interference with energy supply of cells following "uncoupling" of oxidative phosphorylation (12), is it hyperpolarisation and impaired ion-flux as shown for some neuronal cells (22) or is it inhibition of PG-synthesis (2) which was formerly believed to be unequally distributed throughout the body or to show unequal sensitivity to different NSAID in different tissues. There is now evidence that this is not the case (55) which points again to the importance of unequal distribution of NSAID.

Finally, one may ask what may come out of a new concept trying to explain some crucial aspects of the mode of action of NSAID. It appears to us that it may well help to improve existing NSAID by slight chemical modifications as, e.g., "masking" the acidic group to reduce side effects in the stomach. Such efforts have been made, apparently successfully. They will be presented at this meeting and elsewhere (56).Such concept might also allow for detecting other NSAID by focussing experimentation on optimizing their pharmacokinetic behaviour in man. Which in turn could reduce the money and manpower spent for simple screening of the compounds as to their effects on inflammed rat paws (57).

CONCLUSION

It is still an unsettled question by which molecular mechanism(s) NSAID exert their local anti-inflammatory action. Possibly they do it by interferring with a broad variety of processes. Our results together with previous observations (58, 59) clearly show that acidic NSAID reach higher concentrations in inflammed tissue than in most others throughout the body. Only in the absorptive and excretory organs, stomach, small intestine and kidney, can similar high concentrations be seen. This may cause the common side effects of all acidic NSAID in these organs. It may explain why all attempts to develop acidic NSAID without these side effects had only limited success, but it also indicates ways to circumvent some of these side effects with new NSAID.

REFERENCES

(1) R.A. SCHERRER, Introduction to the Chemistry of Anti-inflammatory and Anti-Arthritic Agents, Anti-inflammatory Agents 1, 29-43 (1974), Academic Press, New York.

(2) S.H. FERREIRA and J.R. VANE, New Aspects of the Mode of Action of Nonsteroid Anti-inflammatory Drugs, Ann. Rev. Pharmacol. 14, 57-73 (1974).

(3) A. BENNETT, I.F. STAMFORD and W.A. UNGER, Prostaglandin E_2 and Gastric Acid Secretion in Man, J. Physiol. (Lond.) 229, 349-360 (1973).

(4) J.B. LEE, The Antihypertensive and Natriuretic Endocrine Function of the Kidney: Vascular and Metabolic Mechanisms of the Renal Prostaglandins, Prostaglandins in Cellular Biology, pp.399-450 (1972), Plenum Press, New York.

(5) N.H. GRANT, H.E. ALBURN and C.A. SINGER, Correlation between in vitro and in vivo Models in Anti-inflammatory Drug Studies, Biochem. Pharmacol. 20, 2137-2140 (1971).

(6) T. NARAHSHI, D.T. FRAZIER, T. DEGUCHI, C.A. CLEAVES and M.C. ERNAU, The Active Form of Pentobarbital in Squid Giant Axons, J. Pharmac. exp. Ther. 177, 25-33 (1971).

(7) R.J. FLOWER, Drugs which Inhibit Prostaglandin Biosynthesis Pharmacol. Rev. 26, 33-67 (1974).

(8) A. GOLDSTEIN, L. ARONOW and S.M. KALMAN, Principles of Drug Action, Harper and Row, New York (1969).

(9) H. TERADA, S. MURAOKA and T. FUJITA, Structure-activity Relationships of Fenamic Acids, J. Med. Chem. 17, 330-334 (1974).

(10) W.J. WADDELL and R.G. BATES, Intracellular pH, Physiol. Rev. 49, 285-329 (1969).

(11) N.A. CUMMINGS and G.L. NORDBY, Measurements of Synovial Fluid pH in Normal and Arthritic Knees, Arthritis and Rheumatism 9, 47-56 (1966).

(12) S.S. ADAMS and R. COBB, A Possible Basis for the Anti-inflammatory Activity of Salicylates and other Non-hormonal Anti-rheumatic Drugs, Nature 181, 773-774 (1958).

(13) S. GOLDSTEIN, R. DeMEO and I. SHEMANO, Anti-Inflammatory Activity of 2,4-Dinitrophenol Following Local Administration at the Site of Inflammation, Proc. Soc. Exp. Biol. Med. 128, 980-982 (1968).

(14) W.G. SPECTOR and D.A. WILLOUGHBY, The Pharmacology of Inflammation. Grune and Straton, New York (1968).

(15) H.O.J. COLLIER, New Light on how Aspirin Works, Nature 223, 35-37 (1969).

(16) M.S. STARR and G.B. WEST, The Effect of Bradykinin and Anti-inflammatory Agents on Isolated Arteries, J.Pharm. Pharmacol. 18, 838-843 (1966).

(17) M.J.H. SMITH, P.D. DAWKINS and J.N. McARTHUR, The Relation Between Clinical Anti-inflammatory Activity and the Displacement of L-tryptophan and a Dipeptide from Human Serum in vitro, J. Pharm. Pharmacol. 23, 451-458 (1971).

(18) H.E. PAULUS and M.W. WHITEHOUSE, Nonsteroid Anti-inflammatory Agents, Ann. Rev. Pharmacol. 13, 107-124 (1973).

(19) L.J. IGNARRO and C. COLOMBO, Enzyme Release from Guinea-pig Polymorphonuclear Leucocyte Lysosomes Inhibited in vitro by Anti-inflammatory Drugs, Nature New Biology 239, 155-157 (1972).

(20) A.L. WILLIS, P. DAVISON, P.W. RAMWELL, W.E. BROCKLEHURST, and B. SMITH, Release and Actions of Prostaglandins in Inflammation and Fever: Inhibition by Anti-inflammatory and Antipyretic Drugs, Prostaglandins in Cellular Biology, pp. 227-259 (1972), Plenum Press, New York.

(21) K. BRUNE and M. GLATT, The Avian Microcrystal Arthritis IV. The Impact of Sodium Salicylates, Acetaminophen and Colchicine on Leukocyte Invasion and Enzyme Liberation in vivo, Agents and Actions 4, 101-107 (1974).

(22) H. LEVITAN and J.L. BARKER, Salicylate: A Structure Activity Study of its Effects on Membrane Permeability, Science 176, 1423-1425 (1972).

(23) S.A. KEATS, Discussion, International Symposium on Pain, pp. 517-518, Raven Press, New York (1974).

(24) J.R. VANE, Inhibition of Prostaglandin-synthesis as a Mechanism of Action for Aspirin-like Drugs, Nature New Biology 231, 232-235 (1971).

(25) E.A. HAM, V.J. CIRILLO, M. ZANETTI, T.Y. SHEN and F.A. KUEHL, Studies on the Mode of Action of Non-steroidal Anti-inflammatory Agents, Prostaglandins in Cellular Biology, pp. 345-352, Plenum Press, New York-London (1972).

(26) R.E. LEE, The Influence of Psychotropic Drugs on Prostaglandin Biosynthesis, Prostaglandins 5, 63-68 (1974).

(27) T.W. HARRITY and M.B. GOLDLUST, Anti-complement Effects of Two Anti-inflammatory Agents, Biochem. Pharmacol. 23, 3107-3120 (1974).

(28) M. GLATT, B. PESKAR and K. BRUNE, Leukocytes and Prostaglandins in Acute Inflammation, Experientia 30, 1257-1259 (1974).

(29) P. GRAF, M. GLATT and K. BRUNE, Acidic Non-steroid Anti-inflammatory Drugs Accumulating in Inflammed Tissue, Experientia 31, 951-953 (1975).

(30) C.J. PFEIFFER and J. WEIBEL, The Gastric Mucosal Response to Acetylsalicylic Acid in the Ferret: An Ultrastructural Study, Amer. J. Digest. Dis. 18, 834-846 (1973).

(31) M. JABBARI and L.S. VALBERG, Role of Acid Secretion in Aspirin-induced Gastric Mucosal Injury, Can. med. Ass. J. 102, 178-181 (1970).

(32) W.B. THORSON, D. WESTERN, Y. TANAKA and J.F. MORRISSEY, Aspirin Injury to the Gastris Mucosa. Gastrocamera Observations of the Effect of pH, Arch. int. Med. 121, 499-506 (1968).

(33) G.H. JENNINGS, Causal Influences in Haematemesis and Melaena, Gut 6, 1-13 (1965).

(34) K.D. RAINSFORD, Electronmicroscopic Observations on the Effects of Orally Administered Aspirin and Aspirin-Bicarbonate Mixtures on the Development of Gastric Mucosal Damage in the Rat, Gut 16, 514-527 (1975).

(35) A.M. GOTTSCHALK and R. MENGUY, Influence of Gastric Atrophy on the Susceptibility of the Gastric Mucosa to Injury by Aspirin, Surg. Forum 21, 300-301 (1971).

(36) K.W. ANDERSON, A Study of the Gastric Lesions Induced in Laboratory Animals by Soluble and Buffered Aspirin, Arch. int. Pharmacodyn. 152, 392-403 (1964).

(37) T. GLARBORG-JORGENSEN, E.L. KAPLAN and G.W. PESKIN, Salicylate Effects on Gastric Acid Secretion, Scand. J. clin. Lab. Invest. 33, 31-38 (1974).

(38) K.D. RAINSFORD, The Relationship between Distribution of Salicylates in the Gastrointestinal Tract and Blood and Development of Gastric Damage After Oral Administration of Aspirin and Aspirin-Bicarbonate Mixtures in Rats, Clin. exp. Pharmac. Physiol. 2, 45 (1975).

(39) J.T. SCOTT, A.M. DENMAN and J. DARLING, Renal Irritation Caused by Salicylates, Lancet. I, 344-348 (1963).

(40) A. MORALES and J. STEYN, Papillary Necrosis Following Phenylbutazone Ingestion, Arch. Surg. 103, 420-421 (1971).

(41) O. OLAFSSON, K.R. GUDMUNDSSON and A. BREKKAN, Migraine, Gastritis and Renal Papillary Necrosis. A Syndrome in Chronic Non-Obstructive Pyelonephritis, Acta med. scand. 179, 121-128 (1966).

(42) J.H. SHELLEY, Phenacetin, through the Looking Glass, Clin. Pharmac. Ther. 8, 427-471 (1967).

(43) J.A. ABEL, Analgesic Nephropathy - a Review of the Literature, 1967-1970, Clin Pharmac. Ther. 12, 583-598 (1971).

(44) A.F. MACKLON, A.W. CRAFT, M. THOMPSON and D.N.S. KERR, Aspirin and Analgesic Nephropathy, Brit. med. J. 1, 597-600 (1974).

(45) T. MURRAY and M. GOLDBERG, Analgesic Abuse and Renal Disease, Ann. Rev. Med. 26, 537-550 (1975).

(46) R.S. NANRA and P. KINCAID-SMITH, Papillary Necrosis in Rats Caused by Aspirin and Aspirin-Containing Mixtures, Br. Med. J. 3, 559-561 (1970).

(47) D.L. BOKELMAN, W.J. BAGDON, P.A. MATTIS and P.F. STONIER, Strain-dependent Renal Toxicity of a Nonsteroid Anti-inflammatory Agent, Toxic. appl. Pharmac. 19, 111-124 (1971).

(48) L. ARNOLD, C. COLLINS and G.A. STARMER, Renal and Gastric Lesions after Phenylbutazone and Indomethacin in the Rat, Pathology 6, 303-313 (1974).

(49) J.E. CLAUSEN, Histological Changes in Rabbit Kidneys Induced by Phenacetin and Acetylsalicylic Acid, Lancet, p.123-124 (1964).

(50) L. ARNOLD, C. COLLINS and G.A. STARMER, Renal and Gastric Lesions after Phenylbutazone and Indomethacin in the Rat, Pathology 6, 303-313 (1974).

(51) E.H. WISEMAN and H. REINERT, Anti-inflammatory Drugs and Renal Papillary Necrosis, Agents and Actions 5, 322-325 (1975).

(52) E.G. McQUEEN, Salicylate Toxicology, This Symposium.

(53) W.J. WADELL and T.C. BUTLER, The Distribution and Excretion of Phenobarbital, J. Clin. Invest. 36, 1217-1225 (1957).

(54) J.G. LOMBARDINO, Enolic Acids with Anti-inflammatory Activity, Anti-inflammatory Agents I, 130-158 Academic Press, New York (1974).

(55) S.S. PONG and L. LEVINE, Prostaglandin Synthetase Systems of Rabbit Tissues and their Inhibition by Nonsteroidal Anti-inflammatory Drugs, J. Pharmacol. exp. Ther. 196, 226-230 (1976).

(56) K.D. RAINSFORD and M.W. WHITEHOUSE, Gastric Irritancy of Aspirin and its Analogues: Anti-inflammatory Effects without this Side-effect, J. Pharm. Pharmac., in press (1976).

(57) K. BRUNE, How Aspirin might work: A Pharmacokinetic Approach, Agents and Actions 4, 230-232 (1974).

(58) G. WILHELMI and R. PULVER, Untersuchungen zur Frage eines peripheren Angriffspunktes der Pyrazole bei der antiphlogistischen Wirkung. Drug Res. 5, 221-224 (1955).

(59) A. BENAKIS, G. TSOUKAS and B. GLASSON, Localization of Butylmalonic Acid-mono (1,2-Dephenylhydrazide)-Calcium ^{14}C (Bumadizone-calcium), a new Anti-inflammatory Drug, in Rat and Mice with Wholebody Autoradiography, Drug Res. 23, 1231-1236 (1973).

(60) K.D. RAINSFORD and K. BRUNE, Role of the Parietal Cell in Gastric Damage Induced by Aspirin and Related Drugs: Implication for Safer Therapy, Med. J. Aust. 1, 881-883 (1976).

Acknowledgements:

In order to prepare this manuscript we got permission from the Birkhauser Verlag, Basel and The Australian Medical Publishing Comp., to use tables and figures from our work: P. Graf, M. Glatt and K. Brune: Acidic Nonsteroid Anti-inflammatory Drugs Accumulating in Inflammed Tissue (29).
K. Brune, P. Graf and M. Glatt: Inhibition of Prostaglandin Synthesis in vivo by Nonsteroid Anti-inflammatory Drugs: Evidence for the Importance of Pharmacokinetics (60).
K.D. Rainsford and K. Brune: Role of the Parietal Cell in Gastric Damage Induced by Aspirin and Related Drugs: Implications for Safer Therapy (60).
The cooperation of the Publishers is gratefully acknowledged.

This work was in part supported by grant 3.588-0.75 from the Swiss National Foundation for Scientific Research.

ARE THERE ANTIRHEUMATIC DRUGS WHICH COULD INCREASE THE SYNTHESIS OF PROSTAGLANDINS ?

by J.P. Famaey, J. Fontaine and J. Reuse
Laboratory of Pharmacology, Rheumatology Unit, School of Medicine and Pharmacology, University of Brussels, Brussels, Belgium.

It is well known that most of the classical anti-rheumatic drugs are able to interfere with prostaglandins production (1). The inhibitory effect of aspirin-like drugs on the microsomal prostaglandin-synthetase system has been well documented (1). Mepacrine, an anti-rheumatic antimalarial drug is able to inhibit the phospholipases involved in the production of prostaglandins precursors as well as the anti-inflammatory steroids which have been also described to prevent the prostaglandin-release in entire cell preparations (2). The classical anti-gout agent colchicine, as well as vinblastine, another antitubulin agent, have been shown to increase the prostaglandin synthesis in various in vitro and in vivo models (3,4). Finally the long-acting antirheumatic agents, d-penicillamine and gold salts have been claimed to impair the balance between the supposed pro-inflammatory prostaglandins E and the supposed anti-inflammatory prostaglandins F_α, the gold salts probably increasing the PGE's synthesis and inhibiting the PGF's synthesis (5) and d-penicillamine increasing the well known opposite effect of copper ions (6).

We have used the isolated guinea-pig ileum as a simple and practical in vitro model (7) for testing the ability of various agonists (including some of those involved in the inflammatory reactions) to induce smooth muscle contractions in the presence of classical antirheumatic agents.

We had previously demonstrated that the non-steroidal anti-inflammatory drugs (NSAID) or aspirin-like drugs (e.g. indomethacin, the propionic acids, phenylbutazone and its analogues, bufexamac, alclofenac) are able to inhibit dramatically the isotonic contractions induced by indirect agonists such as nicotine or serotonine as well as the isometric coaxial electrically-induced contractions (7). The direct agonists such as acetylcholine or histamine showed only a slow rate of inhibition in the presence of similar concentrations of NSAID (8). We have also found that a similar effect occurs for some of the most classical anti-inflammatory steroids (e.g. hydrocortisone, prednisone, prednisolone, betamethasone, dexamethasone) (7). By adding a small amount of prostaglandins E_1, E_2 or $F_{2\alpha}$ to the bath, all these inhibitory effects exerted by NSAID or steroids are totally reversed (7). In some cases the observed contractions are even greater than controls (8).

This could be explained by the well documented effects of NSAID and steroids on biological membranes (9). A partial deactivation of ileal smooth muscle membrane receptors to agonists as well as a more important effect at the neuronal membrane level could cause inhibition of the release of endogenous agonists (e.g. acetylcholine) under the influence of indirect agonists.

The reversing effects of prostaglandins could be only due to a non specific sensitization of the ileal smooth muscle to any kind of agonists in the presence of very small amounts of prostaglandins. The argument for such an hypothesis has been presented elsewhere (7) (8).

We have now shown that colchicine (10) and vinblastine (11), the two antitubulin agents known to increase the prostaglandins synthesis can also sensitize the ileal smooth muscle to various agonists (see Table I) as well as d-penicillamine (see Table II) (12) while the gold salts (at least gold thiosulfate) do not seem to have such an effect. Colchicine appears to be the more potent as it is able to act at concentrations of 0.1 and 1 mg/l, while vinblastine and d-penicillamine show an activity only at 10 and 40 mg/l (see fig.1). However, it must be pointed out that, with the exception of colchicine at a concentration of 0.1 mg/l, the observed effect seems to be irreversible. This could be due to the fact that at the concentrations employed these drugs could have some very profound effect on the smooth muscle cells structure. The effect of these drugs on stimulating the synthesis of prostaglandins could be one of the explanations for the observed sensitization. However, it seems likely that the potent ability of antitubulin agents to bind to intracellular proteins (e.g. from microtubules or perhaps from other subcellular structures important for the smooth muscle cells contractility) or to affect membrane proteins, as well as the well known SH binding property of d-penicillamine which is related to its disrupting action on various macromolecules, are other good explanations for the present results.

It is interesting to note that two drugs which have some common clinical uses, namely d-penicillamine and colchicine, in the treatment of systemic sclerosis show some very similar effects in a quite simple _in vitro_ model. However, it is most unlikely that these sensitizing effects have any relationship to the clinical (antirheumatic) properties of these drugs or even with their most classical side effects. Nevertheless this _in vitro_ system could provide a very simple model for testing this kind of antirheumatic agents and the observed results could perhaps allow us to do some clinical extrapolations such as e.g. suggesting the use of vinblastine in the treatment of systemic sclerosis.

REFERENCES

(1) S.H. FERREIRA and J.R. VANE, New Aspects of the Mode of Action of Non-Steroid Anti-Inflammatory Drugs, Ann. Rev. Pharmac., _14_, 57-73 (1974).

(2) R.G. GRYGLEWSKI, Prostaglandins and Prostaglandin synthesis inhibitors in etiology and treatment of inflammation, in : Proceedings of the Sixth International Congress of Pharmacology, Helsinki, Finland, vol. 5, clinical pharmacology (Ed. J. Tuomisto and M.K. Paasonen) pp.151-160 (1975).

(3)   D.R. ROBINSON and R.A. LEVINE, Prostagladin Concentrations in Synovial Fluid in Rheumatic Diseases : Action of Indomethacin and Aspirin, in : Prostaglandin Synthetase Inhibitors : Their Effects on Physiological Functions and Pathological States (Eds. H.J. Robinson and J.R. Vane, Raven Press, New York) pp. 223-228 (1974).

(4)   M. GLATT, P. GRAF and K. BRUNE, Effects of Colchicine in acute inflammation, Experientia, 31, 728 (1975).

(5)   K. STONE, S. MATHER and P. GIBSON, Selective inhibition of prostaglandin biosynthesis by gold salts and phenylbutazone, Prostaglandins, 10, 241-251 (1975).

(6)   I.S. MADDOX, The role of copper in prostaglandins synthesis, Biochim. Biophys. Acta, 306, 74-81 (1973).

(7)   J.P. FAMAEY, J. FONTAINE and J. REUSE, Inhibiting Effects of Morphine, Chloroquine, Non-Steroidal and Steroidal Antiinflammatory Drugs on Electrically-Induced Contractions of Guinea Pig Ileum and Reversing Effects of Prostaglandins, Agents and Actions, 5, 354-358 (1975).

(8)   J.P. FAMAEY, J. FONTAINE and J. REUSE, The effects of various non-steroidal antiinflammatory drugs on cholinergic and histamine-induced contractions of guinea-pig ileum Brit. J. Pharmac., in press (1977).

(9)   J.P. FAMAEY, P.M. BROOKS and W.C. DICK, Biological effects of non-steroidal antiinflammatory drugs, Seminars in Arthritis and Rheumatism, 5, 63-81 (1975).

(10) J.P. FAMAEY, J. FONTAINE and J. REUSE, Smooth muscle sensitization induced by colchicine : is it an in vitro property of antitubulin agents ?, Agents and Actions, in press (1977).

(11) J.P. FAMAEY, J. FONTAINE and J. REUSE, Smooth muscle sensitization induced by Vinblastine, Agents and Actions, 6, 724-727 (1976).

(12) J.P. FAMAEY, J. FONTAINE and J. REUSE, The effects of D-penicillamine on ileal smooth muscle and their possible relationship with prostaglandins, in : Penicillamine research in rheumatoid disease,(Ed. E. Munthe, Fabritius & Sønner, Oslo) pp 50-58 (1976).

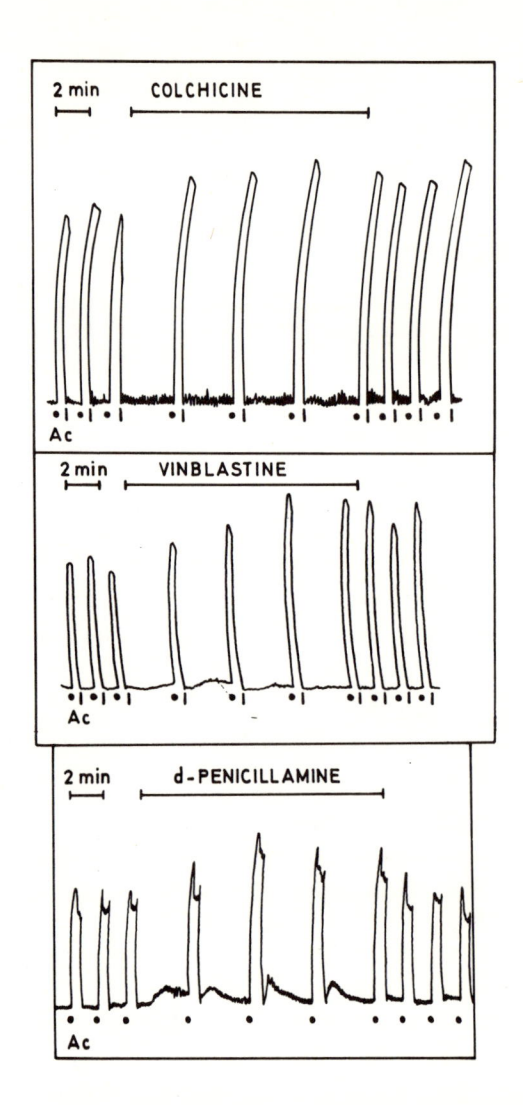

Fig.1. The effects of colchicine(1mg/l), vinblastine(10mg/l) and d-penicillamine(40mg/l) on guinea-pig ileum contractions induced by acetylcholine (Ac) 20 µg/l.

TABLE I

The Sensitizing Effects of Colchicine and Vinblastine on Contraction in the Guinea Pig ileum induced by Various Agonists. From Famaey, Fontaine and Reuse, 1976 (10) (11).

Drugs (mg/l) Agonists	Colchicine 0.1	Colchicine 1	Vinblastine 10	Vinblastine 40
Acetylcholine	119.4 ± 2.7 **	110.4 ± 3.7 *	133.7 ± 5.5 ***	99.4 ± 13.0 N.S.
Histamine	119.9 ± 9.2 N.S.	84.3 ± 7.6 N.S.	114.9 ± 4.5 **	84.9 ± 6.9 N.S.
Nicotine	118.6 ± 3.6 *	99.1 ± 15.0 N.S.	130.1 ± 5.7 ***	93.2 ± 19.3 N.S.
Serotonine	120.9 ± 12.4 N.S.	135.9 ± 6.3 *	112.7 ± 9.2 N.S.	116.9 ± 13.1 N.S.
PGE$_1$	102.4 ± 15.0 N.S.	99.5 ± 10.5 N.S.	110.8 ± 3.0 *	88.9 ± 18.2 N.S.
PGE$_2$	121.2 ± 6.7 *	99.1 ± 17.9 N.S.	-	-
PGF$_{2\alpha}$	-	-	130.6 ± 16.8 *	170.5 ± 15.2 ***

Results are expressed in percentages of control contractions obtained with the same guinea-pig ileum before the addition of the drug; mean ± S.E.M. (n=6, at least).
*P< 0.05 **P< 0.02 ***P< 0.01   N.S. = not significant (Student's "t" test).

TABLE II

The Sensitizing Effect of d-Penicillamine on the Guinea Pig Ileum Contractions Induced by Various Agonists. From Famaey, Fontaine and Reuse, 1976 (12).

Agonists \ concentrations (mg/l)	10	40
Acetylcholine	118.4 ± 4.2 *	144.4 ± 15.0 *
Histamine	95.4 ± 2.4 N.S.	95.3 ± 9.2 N.S.
Nicotine	123.8 ± 10.6 *	134.5 ± 12.3 *
Serotonine	117.0 ± 7.3 N.S.	130.3 ± 12.7 *
PGE_1	114.3 ± 8.9 N.S.	115.0 ± 5.4 *

Results are expressed in percentages of control contractions obtained with the same guinea-pig ileum before the addition of the drugs, mean ± S.E.M. (n=6).
* $P < 0.05$ (Student's "t" test)
N.S. = not significant.

ACTIONS OF ANTI-INFLAMMATORY DRUGS ON SMOOTH MUSCLE.

by D.M.Temple and I.S.McKnight
Department of Pharmacology, University of Sydney, Sydney, N.S.W.
Australia

The ability of anti-inflammatory drugs of the aspirin type to inhibit the biosynthesis of prostaglandins (1) probably explains many of their pharmacological actions. Prostaglandins are synthesized and metabolized in many tissues of the body, including smooth muscle, and it has been postulated, at least in bronchial, uterine and vascular smooth muscle, that they may play a role in the maintenance of muscle tone. In addition to their inhibition of prostaglandin biosynthesis, anti-inflammatory drugs have a direct effect on isolated smooth muscle. The relaxation of isolated human bronchi was induced by treatment with fenamates (2), which blocked the contractile effects of prostaglandin $F_{2\alpha}$ and SRS-A on this tissue.

Anti-inflammatory drugs may affect the prostaglandin system in respiratory tissue many ways. Firstly, prostaglandin biosynthesis and release can be inhibited in lung by indomethacin and aspirin at concentrations in the 10^{-5}M to 10^{-6}M range (1). Secondly, at slightly higher concentration levels, certain anti-inflammatory drugs antagonize the contractile effects of prostaglandin $F_{2\alpha}$ on airways smooth muscle (2). Thirdly, the metabolism of prostaglandins may be inhibited by anti-inflammatory drugs, as has been demonstrated in lung tissue with indomethacin and aspirin (3), and phenylbutazone (4).

Our experiments have utilized isolated preparations of guinea pig trachea (spirally cut) and human bronchus (5). Similar experiments were carried out with isolated strips of rabbit and human myometrium (6). These experiments have shown that the responses of the guinea pig isolated trachea and human bronchus preparations are very similar to each other when either prostaglandin E_1, prostaglandin $F_{2\alpha}$ or flufenamate solutions are added cumulatively to the preparations. For instance, the EC_{50} for relaxation of human bronchus by cumulative addition of flufenamate is 2×10^{-5}M for guinea pig trachea. Figure 1 shows this concentration-dependent relaxation of guinea pig tracheal muscle for a number of anti-inflammatory drugs. It can be seen that the concentration-response curves are approximately parallel, suggesting a common mode of action, but are not parallel with curves produced by the sympathomimetic drugs salbutamol and isoprenaline. The progressive reduction in tone of tracheal muscle may be due to the inhibition of prostaglandin synthesis. Since prostaglandin $F_{2\alpha}$ causes contraction of respiratory smooth muscle and prostaglandins E_1 and E_2 cause relaxation, there seems to be a differential inhibition so that the effect of prostaglandin E may override that of prostaglandin $F_{2\alpha}$. Perhaps prostaglandin $F_{2\alpha}$ biosynthesis is inhibited more than that of E, or the metabolism of prostaglandin E is inhibited more than that of $F_{2\alpha}$.

In isolated preparations of uterine smooth muscle, the contractile effect of added prostaglandin $F_{2\alpha}$ is blocked by the presence of anti-inflammatory drugs. For human myometrium, phenylbutazone antagonized prostaglandin induced contractions of EC_{50} values of 3×10^{-5} M, indomethacin at 4×10^{-5} M and aspirin at 5×10^{-4} M. For rabbit, comparable EC_{50} values obtained for indomethacin were 4×10^{-6} M and for aspirin 4×10^{-4} M (6).

The concentrations of anti-inflammatory drugs which reduce the tone of smooth muscle preparations or which block the actions of added contractile drugs such as prostaglandin $F_{2\alpha}$ in vitro are comparable to the plasma concentrations obtained by these drugs in vivo. It, therefore, seems possible that anti-inflammatory drugs could cause a relaxation of the tone of smooth muscle in vivo.

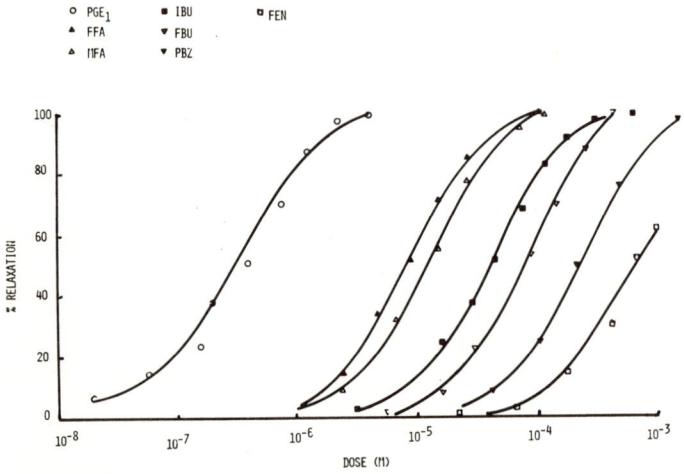

Figure 1.

Log dose response curves for relaxation of the guinea pig isolated spirally cut trachea preparation. Each curve was obtained by cumulative addition of a solution of the drug to a 15 ml organ bath, in which the tracheal strip was suspended in carbogenated Krebs' solution at 37°, and from which changes in tone of the muscle were recorded through a transducer. Responses for each drug were measured as a percentage of the maximum relaxation produced by that drug, and each point represents the mean value from at least three experiments. Anti-inflammatory drugs were dissolved as Na salts. PGE_1 = Prostaglandin E_1, FFA = flufenamate, MFA = mefenamate, IBU = ibuprofen, FBU = flurbiprofen, PBZ = phenylbutazone, FEN = fenclofenac.

REFERENCES

(1) J.R.VANE, Inhibition of Prostaglandin Synthesis as a Mechanism of Action for Aspirin-Like Drugs, Nature (New Biol.) 231, 232-235 (1971).

(2) H.O.J.COLLIER and W.J.F.SWEATMAN, Antagonism by Fenamates of Prostaglandin $F_{2\alpha}$ and of Slow Reacting Substance on Human Bronchial Smooth Muscle, Nature 219, 864-865 (1968).

(3) H.HANSEN, Inhibition by Indomethacin and Aspirin of 15-Hydroxyprostaglandin dehydrogenase in vitro, Prostaglandins 8, 95-105 (1974).

(4) D.J.CRUTCHLEY and P.J.PIPER, Prostaglandin Inactivation in Guinea Pig Lung and its Inhibition, Brit.J.Pharmacol. 52, 197-203 (1974).

(5) I.S.McKNIGHT and D.M.TEMPLE, A Comparison of the Effects of Prostaglandins and Some Relaxant Drugs on Isolated Preparations of Human Bronchus and Guinea Pig Trachea, Clin. Exp.Pharmacol.Physiol., in press (1976).

(6) I.D.SMITH, D.M.TEMPLE and R.P.SHEARMAN, The Antagonism by Anti-inflammatory Analgesics of Prostaglandin $F_{2\alpha}$-induced Contractions of Human and Rabbit Myometrium in vitro, Prostaglandins 10, 41-57 (1975).

SALICYLATES IN RHEUMATOID ARTHRITIS: PHARMACOKINETICS AND ANALGESIC RESPONSE.

by G.G.Graham, G.D. Champion, R.O. Day, A.L.Kaski, L.G.Hills and P.D.Paull
Departments of Clinical Pharmacology and Rheumatology,
St.Vincent's Hospital, Darlinghurst, N.S.W. 2010, Australia.

Two aspects of the clinical pharmacology of aspirin and salicylate will be discussed in this communication. Firstly, we are examining the pharmacokinetics of salicylate in subjects with rheumatoid arthritis. The second part of this communication is a comparison of the activity of aspirin and sodium salicylate in relieving the pain of rheumatoid arthritis.

1. Metabolism of Aspirin:

The results of both studies will be discussed with particular reference to the metabolism of aspirin. Aspirin is rapidly hydrolysed to salicylate and the half life of aspirin is approximately 15 minutes ([1]). Further, only about two-thirds of an oral dose of aspirin is effectively absorbed intact, the remainder being hydrolysed during absorption and first pass through the liver ([2]). The half life of salicylate is dependent upon the body content of the drug and ranges from 3 to about 20 hours ([3]). Limited capacities of the pathways converting salicylate to salicylurate and to salicyl phenolic glucuronide are responsible for the dose dependent-kinetics ([4]). Salicylate is more slowly eliminated than aspirin. Consequently the accumulation of salicylate determines the dosage of the parent drug, aspirin, when this drug is used as an anti-inflammatory agent. However, it should be noted that aspirin acetylates various body constituents ([5]). Such acetylation appears almost irreversible and aspirin could have a prolonged duration of action despite its short half life.

2. Pharmacokinetics of Salicylate in Subjects with Rheumatoid Arthritis:

It is well known that there are considerable intersubject variations in the plasma levels of salicylate during therapy with aspirin. Even when the daily dose of aspirin is as high as 4.8 g, many subjects do not achieve therapeutic levels of salicylate. While there has been no formal study relating suppression of inflammation to the plasma levels of salicylate, considerable clinical evidence indicates that plasma levels in excess of 150 μg/ml are usually required for clinically significant activity ([6]).

We are examining the pharmacokinetics of salicylate both in patients with rheumatoid arthritis and in healthy subjects with the aims of:

(i) The determination of the major factors responsible for the intersubject variations in plasma concentrations of salicylate,

(ii) The definition of a single dose test which may enable the selection of those subjects who should or should not achieve therapeutic levels of salicylate with reasonable daily doses of aspirin. We consider 4.8 g to be a reasonable maximal daily dose of aspirin and generally do not exceed this dose.

Our present pharmacokinetic studies involve the administration of single doses of 1.2 g aspirin and measuring both the plasma concentrations of salicylate and the rate of excretion of salicylate and its metabolites in urine over the next 12 to 24 hours.

Plasma concentration data from 3 individuals are shown in Figure 1. OW and RB were male patients with rheumatoid arthritis while JB was an age matched control subject. The plasma concentrations of salicylate declined more rapidly in OW and RB than in JB and the differences in the 12 hourly values (Cp_{12}) were particularly marked.

TABLE 1

Intersubject variations in the initial plasma half life of salicylate and rates of excretion of salicylate metabolites.

Subject	Initial Half Life (Hours)	Total Salicylate	Salicylurate (SU)	Salicyl Glucurononides (SG)	SU/SG
NM	3.1	97.7	46.2	30.7	1.5
RP	4.0	100	63.0	36.3	1.7
OW	2.9	87.8	50.7	32.5	1.6
SC	5.3	84.4	57.7	21.1	2.7
SG	4.7	69.8	41.2	28.3	1.5
WJ	6.5	61.7	32.7	28.3	1.2
EV	5.4	52.2	28.5	22.0	1.3
DH	5.8	63.1	35.5	21.4	1.7

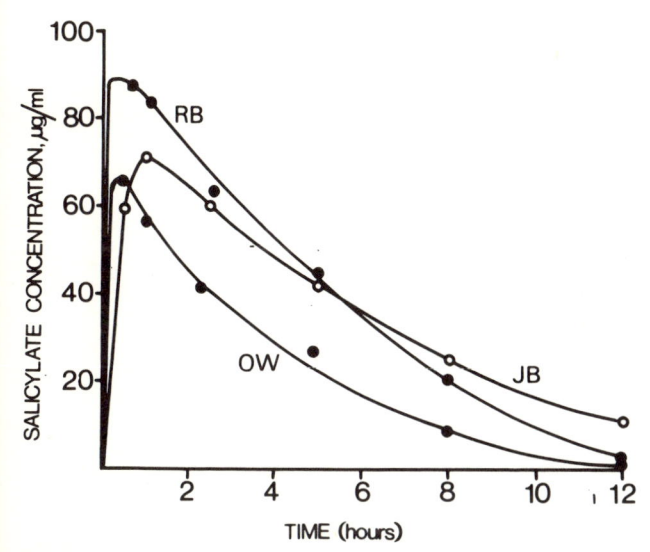

Figure 1.

Plasma concentrations of salicylate following oral dosage with 1.2 g aspirin. OW, RB, patients with rheumatoid arthritis; JB, age matched control subject.

The value of the determination of Cp_{12} was demonstrated from the correlation of Cp_{12} and the plasma levels of salicylate (Cp_{ss}) attained during long term therapy with approximately 4.8 g aspirin per day. Most subjects were dosed with 4.8 g aspirin (16 soluble aspirin tablets) while some received either 4.55 or 5.2 g aspirin per day (7 or 8 enteric coated tablets each containing 650 mg aspirin). The correlation was highly significant (r = 0.83, p < 0.001) (7). In general, those subjects with Cp_{12} values below 5 µg/ml did not attain Cp_{ss} values above 150 µg/ml. Most of these subjects were also being dosed with oral corticosteroids. Conversely, most patients with Cp_{12} values above 10 µg/ml did attain Cp_{ss} values above 150 µg/ml. Only 1 of 7 subjects dosed with oral corticosteroids achieved Cp_{ss} values above 150 µg/ml. Thus, the single dose test appears useful in improving the clinical usage of aspirin by selecting out those subjects who probably will not achieve effective levels of salicylate with reasonable daily doses of aspirin.

The influence of corticosteroids on the plasma levels of salicylate is of interest as combination therapy with corticosteroids and aspirin is frequently used in patients with severe rheumatoid arthritis. Klinenberg and Miller (8) showed that corticosteroids decreased plasma levels of salicylate during long term administration of aspirin. Their clearance studies indicated that corticosteroids increased the urinary excretion of salicylate. However, their assay procedures did not distinguish between salicylate and its major metabolite, salicylurate. In patients treated with corticosteroids in our studies, the excretion of unconjugated salicylate accounted for less than 15 percent of the dose of aspirin ($\underline{7}$). Consequently, corticosteroids must increase the rate of metabolism of salicylate rather than its rate of urinary excretion.

Marked intersubject variations were also noted in the volume of distribution of salicylate. The volumes of distribution ranged from 7.9 to 17.9 litres and correlated rather poorly with body weight (r = 0.51), although the correlation was significant (p < 0.01) ($\underline{7}$). Analysis of the plasma level data indicated that the lower Cp_{12} values in the patients treated with corticosteroids were due to a faster rate of metabolism and not to a higher volume of distribution.

We have also observed intersubject variations in the rates of urinary excretion of salicylate metabolites. As yet, we have only measured the rates of urinary excretion of the total salicylate, salicylurate, total salicyl glucuronides and unconjugated salicylate. Table 1 shows the rates of excretion of the salicyl conjugates from 2 to 8 hours after dosage. During this period, plasma concentrations and total body levels of salicylate declined approximately linearly with time (zero order phase) due to limited capacity of salicylurate and salicyl phenolic glucuronide formation ($\underline{4}$). As may be anticipated, there was a negative correlation (r = 0.77, p < 0.05) between the rate of excretion of total salicylate and the initial half life of salicylate from analysis of the plasma concentration data (Table 1). However, the relative rates of excretion of salicylurate and total salicyl glucuronides did not correlate with the rate of excretion of total salicylate. Apart from subject SC, the ratio of the rate of excretion of salicylurate to the rate of excretion of total glucuronides only ranged from 1.2 to 1.7. Similar results were obtained by Gibson and co-workers ($\underline{9}$) who measured the excretion of the various metabolites of salicylate during long term dosage with aspirin. These investigators found that the ratios of the metabolites in urine were relatively constant and did not correlate with the steady state levels of salicylate.

Our working hypothesis to explain the limited relative rates of formation of metabolites is that the uptake of salicylate by the liver is a rate limiting factor in its metabolism. More detailed pharmacokinetic studies are in progress, in particular studies on the kinetics of elimination of the individual glucuronides after single doses of aspirin, to test our hypothesis.

3. Comparative Analgesic Activity of Aspirin and Salicylate:

In experimental pain, aspirin is a more potent analgesic than sodium salicylate ($\underline{10}$). However, the comparative activity of the two compounds has been little studied in clinical pain, although it is widely assumed that aspirin is the more potent agent in the relief of clinical pain. Our studies were designed to evaluate the relative activity of aspirin and sodium salicylate in the relief of the pain of rheumatoid arthritis. However, aspirin and sodium salicylate are also anti-inflammatory agents and some suppression of inflammation by be partly responsible for their analgesic activity in rheumatoid arthritis.

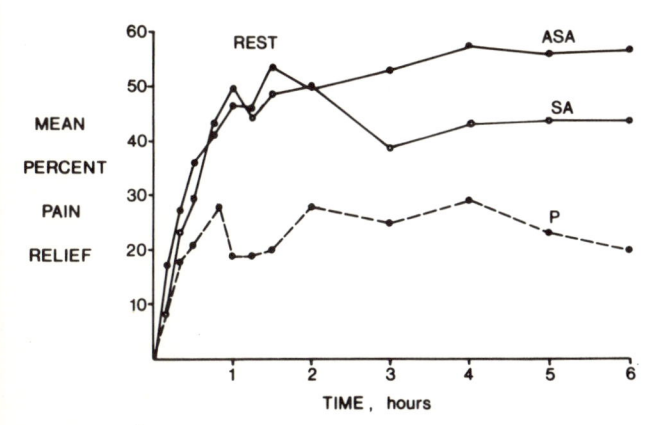

Figure 2.

Mean percent pain relief following dosage with 1.2 g aspirin (ASA), 1.07 g sodium salicylate (SA) and placebo (P).

Analgesic tests have been conducted in 21 subjects with rheumatoid arthritis. Each patient received three treatments at least 24 hours apart, namely solutions of 1.2 g aspirin, its molar equivalent of sodium salicylate (1.07 g) and placebo. Each preparation contained an excess of 0.5 g sodium bicarbonate and was prepared immediately before dosage. The analgesic trials were double blind and the treatment order was according to a randomised block design. Pain was assessed by a visual analogue scale when the patients were at rest and also during brief exercise. The use of a visual analogue scale allows a simple and sensitive measurement of pain (11).

When the patients were at rest, all 3 treatments decreased the pain score relative to the initial pain score (Figure 2). The percent pain relief produced by both drugs was not significantly different from placebo until 45 minutes after dosage. Thereafter, the analgesic effects of both drugs were significant different from placebo ($p \leq 0.05$). There was a trend towards less sustained analgesia with sodium salicylate but the percent pain relief produced by the two drugs was not significantly different at any time.

Similar results were obtained when pain was measured during brief exercise, although the absolute pain scores were higher. Pain scores on exercise and at rest correlated closely ($r = 0.84$, $p \leq 0.001$).

Considering these comparative analgesic activities and the rapid hydrolysis of aspirin in vivo, the effect of aspirin in the relief of the pain of rheumatoid arthritis appears largely due to its metabolite, salicylate.

REFERENCES

(1) M.ROWLAND and S.RIEGELMAN, Pharmacokinetics of Acetylsalicylic Acid and Salicylic Acid after Intravenous Administration in Man, J.Pharm.Sci. 57, 1313-1319 (1968).

(2) M.ROWLAND, S.RIEGELMAN, P.A.HARRIS and S.D.SHOLKOFF, Absorption Kinetics of Aspirin in Man Following Oral Administration of an Aqueous Solution, J.Pharm. Sci. 61, 379-385 (1972).

(3) G.LEVY, Pharmacokinetics of Salicylate Elimination in Man, J.Pharm.Sci. 54, 959-967 (1965).

(4) G.LEVY, T.TSUCHIYA and L.P.AMSEL, Limited Capacity for Salicyl Phenolic Glucuronide Formation and its Effect on the Kinetics of Salicylate Elimination in Man, Clin.Pharmacol.Ther. 13, 258-268 (1972).

(5) R.N.PINCKARD, D. HAWKINS and R.S.FARR, In Vitro Acetylation of Plasma Proteins, Enzymes and DNA by Aspirin. Nature 219, 68-69 (1968).

(6) J. KOCH-WESER, Serum Drug Concentrations as Therapeutic Guides, New Engl.J.Med. 287, 227-231 (1972).

(7) G.GRAHAM, G.D.CHAMPION, R.O.DAY and P.PAULL, Patterns of Plasma Levels and Urinary Excretion of Salicylate in Subjects with Rheumatoid Arthritis, Submitted for Publication.

(8) J.R.KLINENBERG and F.MILLER, Effect of Corticosteroids on Blood Salicylate Concentration, J.Amer.Med.Assoc. 194, 601-604 (1965).

(9) T. GIBSON, G.ZAPHIROPOULOS, J.GROVE, B.WIDDOP and D.BERRY, Kinetics of Salicylate Metabolism, Brit.J.Clin.Pharmacol. 2, 233-238 (1975).

(10) H.O.J.COLLIER, A Pharmacological Analysis of Aspirin, Adv. Pharmacol.Chemother. 7, 333 (1969).

(11) C.R.B.JOYCE, D.W.ZUTSHI, V.HRUBES and R.M.MASON, Comparison of Fixed Interval and Visual Analogue Scales for Chronic Pain, Europ.J.Clin.Pharmacol. 8, 415-420 (1975).

ALTERNATIVES TO ASPIRIN, DERIVED FROM BIOLOGICAL SOURCES

by M.W. Whitehouse, K.D. Rainsford*, N.G. Ardlie, I.G. Young and K. Brune**

Department of Experimental Pathology, Clinical Science and Biochemistry, John Curtin School of Medical Research, The Australian National University, Canberra, Australia 2600; *Department of Biochemistry, University of Tasmania, Hobart, Australia 7001; ** Department of Pharmacology, Biozentrum, University of Basel, Switzerland 4056.

This Symposium is part of a very much larger, international meeting that has taken as its overall theme the topics of Energy, Food and Population. It is therefore a suitable occasion to try to take stock of our drug resources and to peer into the future to see if the traditional sources of raw materials for making the "aspirins" we need will necessarily suffice for the future. At present, aspirin (2-acetoxybenzoic acid) is mainly derived from the following sequence of chemical reactions:

   Fossil Fuel -> Phenol -> Salicylic Acid -> Aspirin.

The day may come when coal and petroleum will no longer be economic raw materials for this purpose, particularly if aspirin, or its substitutes, is/are still to be available as cheap, high volume medication for the masses.

Ortho-carbonyl phenols are found in many plants, produced by certain microorganisms and have even been detected in animal sources (Fig. 1-3 and Table 1). The pathways of salicylate biogenesis in different species are outlined in figures 2-4 (see also refs 2,7). It can be seen that the biological synthesis of large quantities of salicylates could be achieved by selecting suitable strains or mutants with high synthetic capacity. This raises the question as to whether a microbial fermentation process or a field crop could become a significant future source of supply for aspirin or one of its therapeutic alternatives. In this paper we shall consider the therapeutic and side effects of some naturally occurring salicylates and some related derivatives in order to see what alternatives to aspirin are open to us to exploit in the future.

1. CONCERNING THE MERITS OF THE NATURAL SALICYLATES AS ALTERNATIVES TO ASPIRIN

(i) Potential Therapeutic Actions and Side-Effects:

   Salicyl alcohol (saligenin) is of considerable historical interest as the natural salicylate that was much used for the relief of ague and fever and an early source of sali-

Naturally occurring SALICYLATES.
(Plants, microorganisms).

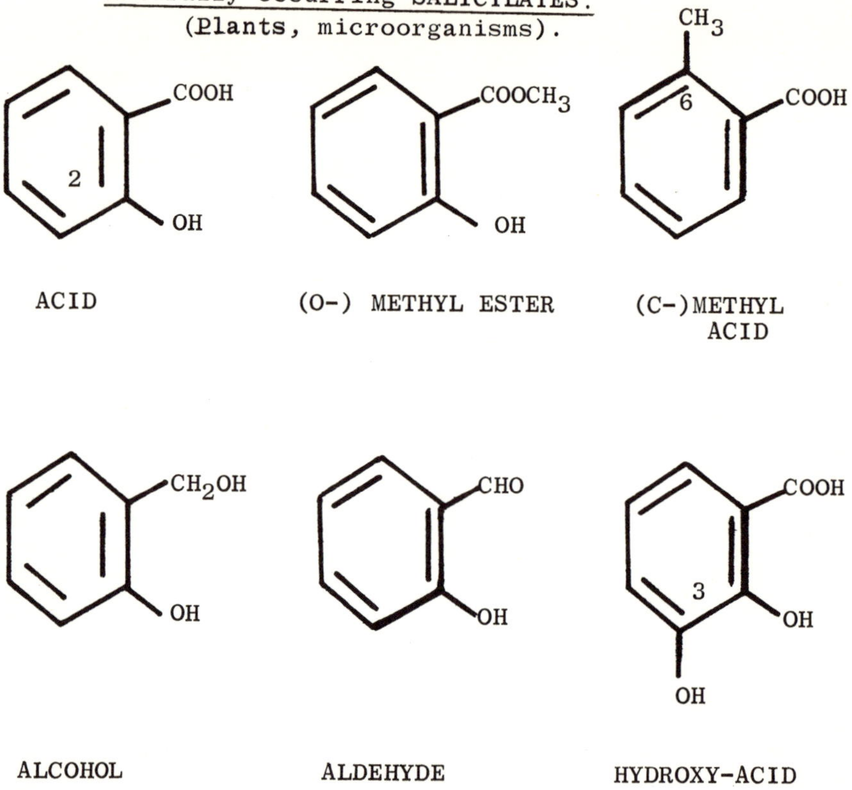

ACID	(O-) METHYL ESTER	(C-) METHYL ACID
ALCOHOL	ALDEHYDE	HYDROXY-ACID

Occur as free PHENOLS

PHENOLIC GLYCOSIDES

cylic acid (1). However, we shall not consider it further here since it failed to show anti-inflammatory activity in an acute assay. For the same reason we also exclude 2 other natural salicylates, gentisic acid and 2-hydroxy-acetophenone (table 2) as potential alternatives to aspirin.

Salicylaldehyde is also of historical interest as being the material obtained from <u>Spirea</u>, from which salicylic acid is prepared in pure form (1). This is commemorated in the German name for Salicylic acid, "Spirsäure". (The name "aspirin" was in turn derived from a = acetyl, spir = salicylic acid). Although the free aldehyde is effective in anti-oedemic assays (table 2) it is not effective in analgesic assays (table 3). Also it is too caustic to the stomach to be considered a useful drug (table 2). In acute assays, its non-irritant dimer has so little activity that it must also be discounted (tables 2 and 3). Salicylaldehyde triacetate certainly shows anti-inflammatory activity (table 2) and is not a stomach irritant (table 2) but its

TABLE 1. Some natural products related to 2-hydroxybenzoic (salicylic) acid.

Compound	Occurs as	in	ref.
1. 2-Hydroxybenzyl (salicyl) Alcohol = Saligenin	Phenolic glucose (salicilin)	Poplar, willow bark Willow leaves	1,2
2. Salicylaldehyde	Free phenol Glucoside (helicin)	Spirea	1,2
3. Salicylic acid	Free acid Mycobactins	Mycobyct. smegmatis (Fe-deficient media)	2
4. Methyl Salicylate	Aglycone Various glycosides Methyl ester	Oils of Wintergreen, Gaultheria, Teaberry Many botanical orders	1,2
5. 6-Methylsalicylic acid	Free acid. Mycobactins	Ponerine ants (alarm pheromone) Penicillium griseofulrum Mycobact. fortuitum (H)	3
6. 2,3-Dihydroxybenzoic acid (Pyrocatechuic acid)	3-O-glucoside N-(lysine, glycine or serine) conjugates e.g. enterocholin	Periwinkle leaves Bacteria (Fe-deficient media)	2 4 2,5
7. 2-Hydroxyacetophenone	Free phenol	Chione glabra (West Indies)	6

SALICYLATE BIOGENESIS : SHIKIMATE PATHWAY.

low analgesic activity in the mouse writhing assay (table 3) and the present lack of any rich source of salicylaldehyde itself, would seem to eliminate simple aldehyde derivatives as possible alternatives to aspirin at this time.

6-Methylsalicylic acid is not at present readily available either although it can be produced in microorganisms. It may be of some future value since it is not ulcerogenic in

TABLE 2: Anti-inflammatory (oedema) and Ulcerogenic activity of some naturally accurring salicylates and the related phenyl acetates.

Phenol	Oedema Δ mean±S.E.	L.I.	Phenyl acetate	Oedema Δ mean±S.E.	L.I.
None [a]	2.45±0.15	0	-	-	-
Salicylic acid	1.05±1.10	11	Aspirin	1.00±0.15	38
Methyl salicylate	1.30±0.15	0	Aspirin methyl ester	1.20±0.10	0
Salicylaldehyde derivs:					
- monomer	1.35±0.15	0[b]	Monacetate	1.25±0.15	[b]
- dimer	2.00±0.10	0	Triacetate	1.55±0.05	0
*1,2-Dihydroxybenzene	2.15±0.30	35	Diacetate	2.65±0.10	0
2,3-Dihydroxygenzoic acid	1.30±0.10	7	Diacetate	1.15±0.20	0
*Methyl salicylic acid	1.25±0.10	0	3-Methyl aspirin	2.05±0.10	7
6-Methyl salicylic acid	1.45±0.15	0	6-Methyl aspirin	1.50±0.15	29
2-Hydroxyacetophenone	2.55±0.10	0	Monoacetate	2.60±0.15	0

Footnotes:
(1) Anti-inflammatory activity measured by foot oedema induced with 1.0mg Na carrageenan in 0.1ml saline in unstarved male Wistar rats (180-250g) dosed orally with 1.1 moles test compound (= 200 mg/kg aspirin suspended in 1% gum acacia). Δ = mean increase paw thickness (mm) ± S.E. is the oedema quantitation of one rear paw at 2 hrs compared with the other in which 0.1ml saline had been injected. Compounds that failed to disperse well in acacia (e.g. liquids) were also assayed after administration in 1% v/v Tween 20-1% acacia. Each set of experiments included groups of animals given aspirin or salicylic acid as the standard drugs.

(2) Gastric irritancy determined as lesion index (L.I.) (12) in male or female starved (20 hr) rats 180-220g given 100 mg/kg test compound and subjected to cold stress at

$-15°$ for 45 mins. (12). Each set of experiments included a group of animals given aspirin as the reference ulcerogen. Compounds prefixed with asterisks included for reference purposes.

(3) $\underline{a}$ = control i.e. no drug but only vehicle (acacia ± Tween 20).

(4) $\underline{b}$ = these aldehydes = gastric irritants but not ulcerogens like aspirin.

(5) Details of compounds shown in appendix 1.

---

rats in contrast to salicylic acid (table 2). 6-Methylaspirin is like aspirin itself a potent ulcerogen (table 2) but nonetheless it too may offer some advantage over aspirin since it does not appreciably affect platelet aggregation (table 5). A clinical trial of 6-methyl aspirin in patients with bleeding disorders would be of some interest especially if this drug were used as a slow or enteric release formulation (to prevent gastric irritancy). However, if 6-methylaspirin has a half life in man no greater than that of aspirin (and salicylic acid), there may be little merit in using it if its intrinsic pharmacological activity is actually lower than that of aspirin itself (tables 2 and 3).

Methyl salicylate is surprisingly innocuous amongst phenolic compounds and is certainly not a gastric ulcerogen in contrast to salicylic acid, although it is irritating when swallowed. Likewise its acetyl derivative, aspirin methyl ester is an attractive alternative to aspirin being non-ulcerogenic (14; table 2).

2-3-Dihydroxybenzoic acid (DHB) itself has certain undesirable properties. It is ulcerogenic, though far less so than either salicylic acid or catechol (1,2-dihydroxybenzene). It is unstable in aqueous solutions, being oxidized to 3-carboxyorthoquinone and polymers are formed therefrom. It is less potent than salicylic acid as an uncoupling agent (15,16). Nevertheless, it is well tolerated in rats and rabbits and has been used in clinical studies as an iron-chelator (17). Its potential availability is discussed further below. 2,3-Diacetoxybencoic acid (DAB) is, by contrast, stable in solution, has anti-inflammatory and analgesic activity comparable to salicylic acid (15) and is not ulcerogenic (tables 2 and 3). We have established that DAB is absorbed rapidly after oral administration giving significant blood levels within 15 minutes in rats. DAB is less toxic than aspirin to arthritic animals but this may perhaps reflect the shorter half life (of both DAB and DHB)

TABLE 3: **Analgesic activity in mice: inhibition of writhing elicitial with acetylcholine (Ac.Ch).**

Number of writhes recorded for each mouse (groups of 4-8, 10 min after i.p. Ac.Ch.)

Drug dose test compound Av.writhes/mouse	0	<3	<5	100mg/kg >5	0	<3	<5	200mg/kg >5
None			+					
Salicylic acid		+				+		
Aspirin	+							
Methyl salicylate			+		+		+	
Salicylaldehyde			+					+
Salicylaldehyde triacetate			+				+	+
6-Methylsalicylic acid			+				+	
6-Methyl aspirin			+				+	
2,3-Dihydroxybenzoic acid		+					+	
2,3-Diacetoxybenzoic acid		+					+	
Salicylamide		+				+		

Analgesic activity was measured in groups of 4-8 outbred mice (25-40g) by the writhing induced with 3.2mg/kg acetylcholine chloride (Ac.Ch.) in 0.25ml saline injected intraperitoneally 30 min after administration an oral dose of the test compound in acacia. The total number of writhes over the subsequent 10 min was recorded for each mouse. Details of compounds in Appendix 1.

SALICYLATE BIOGENESIS :

From Phenylalanine (plants).

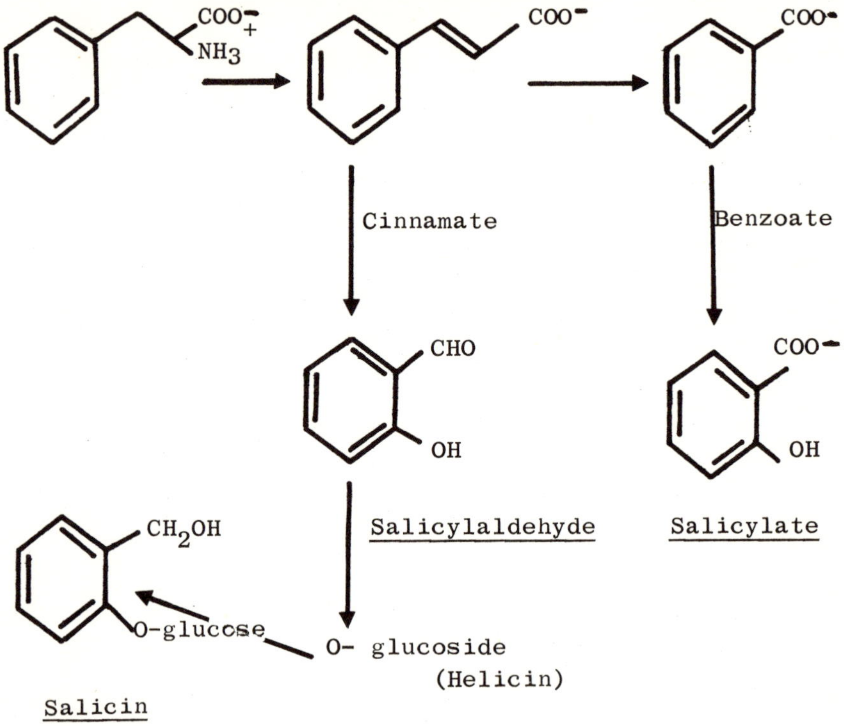

COO-
NH3
COO-
COO-
Cinnamate
Benzoate
CHO
OH
COO-
OH
Salicylaldehyde
Salicylate
CH2OH
O-glucose
O- glucoside
(Helicin)
Salicin

in the plasma. Like aspirin, DAB inhibits platelet aggregation (table 5); a property which may be of some value in certain types of clinical inflammation but perhaps not in others.

In the therapeutic assay for anti-arthritic activity (i.e. against established arthritis, we have observed that the effective dose of aspirin (250mg/kg t.i.d.) is lethal to more than 70% of the animals (i.e. after 12 doses). Equivalent doses of 2,3-diacetoxybenzoic acid were found to be equally as effective as aspirin but notably less toxic in that no animals died. (Efficacy in these studies was determined by 50% reduction in the (increased) arthritic score, compared with that of vehicle-dosed control arthritic animals, observed from the time of initial dosing to 4,5 days later).

We have found that the following compounds were virtually inactive in assays of anti-inflammatory activity: benzoic acid, the 2,4- and 2,6-dihydroxybenzoic acids, the 4- or 5-methylsalicylic acids and the 3,4- or 4,5-dimethyl aspirins. Thus 6-methyl salicylic acid and 2,3-dihydroxybenzoic acid, which occur in nature, are the most active isomers as regards potential anti-inflammatory activity.

It is of interest that the four most important agents inhibiting the platelet release reaction are aspirin, aspirin methyl ester, 2,3-diacetoxybenzoic acid; and 2,3-dihydroxybenzoic acid. These data (table 5) indicate that an acetoxy group in the two position of the benzene ring is important in determining the inhibitory action of aspirin and its analogues, but agents without an acetoxy radical can still be inhibitory. Our results support the conclusions of Mills and coworkers (18). The results of table 5 show that salicylic acid does not inhibit both the second phase of ADP-induced aggregation and collagen-induced aggregation. Methyl salicylate also failed to inhibit collagen-induced aggregation but did inhibit second phase aggregation caused by ADP.

TABLE 4: Effects of Salicylates on Prostaglandin Content in Inflammed Tissue.

Drugs[a]	Dose (mg/kg)	Number of animals	PGF_2 ng/ml[b] mean ±S.D.	PGE_2 ng/ml mean ±S.D.
DMSO (Vehicle)	500	6	7.2±2.0	21.6±10.7
Indomethacin	5	6	0.7±0.5**	0.5± 0.7**
Aspirin	150	6	1.0±1.3**	2.4± 2.9**
Aspirin methyl ester	150	6	4.6±2.7*	15.6± 8.7
2,3-Dihydroxy-benzoic acid	150	4	1.8±0.6**	1.9±0.5**

a) Drugs were dissolved in DMSO so that each animal received the same amount of DMSO (500 mg/kg) which was shown to be without influence on PG-concentration in our experimental system.

b) PG's were assayed in joint fluids by a direct radioimmunoassay as described previously (25). Joint fluids were obtained by washing the intertarsal joints of chickens (2 kg body weight) with 0.5 ml saline one hour after the injection of 0.3 ml of 4% (w/v) suspension of urate crystals, i.e. 2 hours after i.m. administration of the drugs. ** $p < 0.01$; * $p < 0.05$.

Inhibition of the platelet release reaction by aspirin accounts for its effects on ADP and collagen-induced aggregation, and it is likely that the observed effects of the other anti-inflammatory drugs examined here can also be explained by inhibition of the platelet release reaction.

Reduced platelet aggregation, although not primarily causing gastroduodenal bleeding, may contribute to its magnitude and persistence (19). The _in vitro_ studies show that some aspirin analogues with anti-inflammatory activity are less

TABLE 5: Effect of some Salicylate Derivatives on Platelet Aggregation.

Agent	Minimum concentration of agent to completely inhibit aggregation by	
	ADP	Collagen
Aspirin	4.2×10^{-5}M	7.7×10^{-5}M
Methyl Salicylate	4.2×10^{-3}M(100)	No inhibition
Salicylic Acid	No inhibition	No inhibition
Salicylaldehyde	4.2×10^{-3}M(100)	4.2×10^{-3}M(55)
2-acetoxybenzaldehyde	4.2×10^{-3}M(100)	7.7×10^{-3}M(100)
6-methyl Aspirin	7.7×10^{-3}M(183)	7.7×10^{-3}M(100)
2,3-Diacetoxybenzoic acid	2.1×10^{-4}M(5)	7.7×10^{-4}M(10)
2,3-Dihydroxybenzoic acid	8.3×10^{-4}M(20)	3.8×10^{-3}M(49)

Each drug was tested on platelet rich plasma (PRP) from 14 healthy blood donors and male colleagues who had not taken aspirin or any other medication for at least 2 weeks before venepuncture. Blood was mixed with 0.1M trisodium citrate in a ratio of 9:1 (v/v), and (PRP) was obtained by centrifugation the blood. To demonstrate the effects of each anti-inflammatory agent the following concentrations of stock solutions were prepared: 500, 250, 100, 50, 25, 10, 5, 2.5 and 1×10^{-4}M. Platelet aggregation was studied at 37°C by the turbidometric method of Born (20) as modified by Mustard and co-workers (21) using a Payton dual channel Aggregation Module. ADP (Sigma, St. Louis, U.S.A.) was prepared in modified Tyrode's solution before use (22). Collagen was an acid-soluble preparation of achilles tendon obtained from Sigma, St. Louis, U.S.A. Concentrations of aggregating agents (ADP or collagen) used were the minimum necessary to produce maximal aggregation. When higher concentrations were used, higher salicylate concentrations were needed to inhibit aggregation. Figures in parentheses represent the ratio between maximum concentration of agent and minimum aspirin concentration.

6- METHYL SALICYLIC ACID BIOGENESIS :
POLYKETIDE PATHWAY.

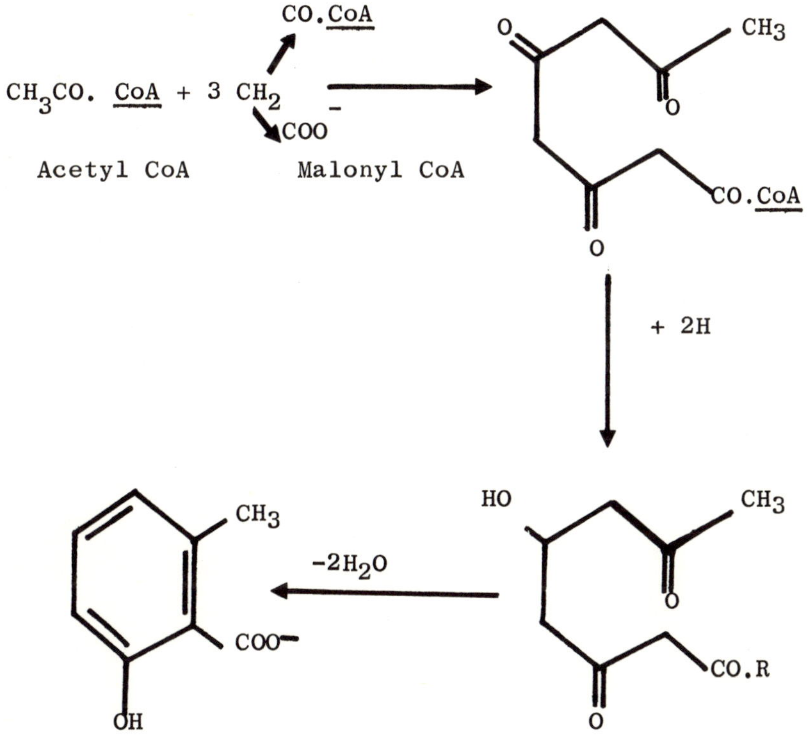

CH$_3$CO. CoA + 3 CH$_2$(COO$^-$)(CO.CoA) ⟶ [triketide intermediate with CO.CoA]

Acetyl CoA   Malonyl CoA

+ 2H ↓

6-M.S.A. ⟵ $-2H_2O$ [reduced intermediate with CO.R]

potent inhibitors of platelet function than aspirin. One would therefore predict less bleeding with these analogues, and indeed their decreased inhibition of platelet function may in part explain why these agents are less ulcerogenic.

2. **Preparation of 2,3-dihydroxybenzoic acid and salicylic acid using microorganisms**

As previously summarized in Table 1, 2,3-dihydroxybenzoic acid is a precursor of the natural iron carrier, enterochelin, which is produced by Escherichia coli and other enteric bacteria. Similarly, salicylic acid and 6-methylsalicylic acid are precursors of the iron-binding mycobactins formed by the mycobacteria. One way of producing these compounds by fermentation is to use mutants which accumulate them owing to metabolic blocks in the respective biosynthetic pathways. Suitable mutants of E. coli which accumulate 2,3-

dihydroxybenzoic acid are already available (23) and presumably analogous mutants of various mycobacteria could also be isolated for the preparation of the salicylic acids.

An alternative approach would be to use the mutant strain of Aerobacter aerogenes which accumulates chorisimic acid (see fig. 2) in good yield (24). The chorisimic acid could then be converted to salicylic acid or 2,3-dihydroxybenzoic acid using the appropriate bacterial enzymes in an immobilized form suitable for large scale preparation.

While the approaches outlined above are perfectly feasible, it is difficult to assess their economic viability. This would depend on the success in developing high-yielding procedures and, at any particular time, on how great the demand was for alternative aspirins in relation to long- or short-term shortages of coal and petroleum.

In conclusion, it appears that of the naturally occurring salicylates or derivatives 2,3-diacetoxybenzoic acid offers considerable promise as a drug with full anti-inflammatory and other therapeutic activities of aspirin but without the gastric ulcerogenicity found with aspirin. This effect may be explained by thoughts outlined in a previous paper (26). Also, 2,3-diacetoxybenzoic acid could be produced by acetylation of 2,3-dihydroxybenzoic acid derived from bacterial sources (by afermentation process). The potential uses for 6-methyl salicylic acid and some other salicylates from natural sources have been discussed.

ABSTRACT

We have examined the "aspirin-like" properties of four naturally occurring salicylic acid (SA) derivatives: methyl salicylate, salicylaldehyde, 2,3-dihydroxybenzoic acid (DHB) and 6-methyl-salicylic acid (6-MSA), together with their respective O-acetyl derivatives in in vivo assay systems employing mice, rats, chickens, pigs, human subjects and an in vitro assay using blood platelets. The principal findings were:

1) 6-MSA, 6-methylaspirin, salicylaldehyde and its acetyl derivatives are all less potent analgesics and anti-inflammatory agents in rodents than SA and aspirin.

2) Methyl salicylate, DHB, aspirin methyl ester (AME) and 2,3-diacetoxybenzoic acid (DAB) exhibit significant anti-inflammatory and anti-arthritic properties in rodents comparable with SA and aspirin..

3) AME and DAB significantly inhibit human platelet aggregation in vitro and erythema development in vivo, comparable to aspirin.

4) 6-MSA, methyl salicylate, AME and DAB are not gastric ulcerogens in stressed rats and pigs whereas DHB, SA, 6-methyl-aspirin and aspirin are weak, modest, potent and very potent ulcerogens respectively in these two animals.

5) The effect of aspirin, DHB and AME in lowering the PG-concentrations in inflammed tissue was investigated. Aspirin and DHB were highly active whilst AME showed moderate but significant effects in the same dose range.

6) The possibility of using microorganisms to produce DHB and other salicylates is considered.

REFERENCES

(1) M. GROSS and L.A. GREENBERG, The Salicylates: A Critical Bibliographic Review., Hillhouse Press, New Haven 1948, pp. 380.

(2) E. HASLAM, The Shikimate Pathway, London Butterworths, 1974, pp. 316.

(3) R.M. DUFFIELD and M.S. BLUM, Methyl 6-methylsalicylate: Identification and Function in a Ponerine Ant, Experientia, 31, 466 (only).

(4) F.E. KING, J.H. GILKS and M.W. PARTRIDGE, A Glycosidic Constituent of Vinca Minor and its Identification as 3-B-D-Glucosyloxy-2-Hydroxybenzoic Acid, J. Chem. Soc. 4206-4215 (1955).

(5) H. ROSENBERG and I.G. YOUNG, Iron Transport in the Enteric Bacteria, in Microbial Iron Metabolism, ed. J.B. Neilands, Academic Press N.Y. 67-82 (1974).

(6) W.R. DUNSTAN and T.A. HENRY, Occurrence of Orthohydroxy-Acetophenone in the Volatile Oil of Chione-Glabra, J. Chem. Soc. 75, 66-71 (1899).

(7) I.G. YOUNG, F. GIBSON and C.G. MacDONALD, Enzymic and Nonenzymic Transformations of Chorismic acid and related Cyclohexadienes, Biochem. Biophys. Acta, 192, 62-72 (1969).

(8) M.A. STAHMANN, I. WOLFF and K.P. LINK, The Synthesis of 4-hydroxycoumarins, J. Am. Chem. Soc. 65, 2285-2287 (1941).

(9) A. NEUBERGER, Synthesis and Resolution of 2,5-Dihydroxy-phenyl Alanine, Biochem. J. 43, 599-605 (1948).

(10) R. ADAMS, M.F. WOGLER and C.W. KREGER, The Structure of Disalicyl Aldehyde, J. Am. Chem. Soc. 44, 1127-1133 (1922).

(11) M. SIMOKORIYAMA, Ueber der Bildung der partiellen Acetate von Flavonen, Anthrochinonen und dergleichen Verbindungen, Bull. Chem. Soc. Japan, 16, 284-291 (1941).

(12) K.D. RAINSFORD, A Synergistic Interaction between Aspirin or other Non-steroidal Anti-inflammatory Agents, and Stress which Produces Severe Gastric Mucosal Damage in Rats and Pigs, Agents and Actions, 5, 553-558 (1975).

(13) S.S. ADAMS, K.F. McCULLOUGH and J.S. NICHOLSON, The Pharmacological Properties of Ibuprofen, an Anti-inflammatory, Analgesic and Antipyretic Agent, Arch. Int. Pharmacodyn. 178, 115-120 (1969).

(14) K.D. RAINSFORD and M.W. WHITEHOUSE, Gastric Irritancy of Aspirin and its Analogues: Anti-inflammatory Activity without this Side-effect, J. Pharm. Pharmac. in the press (1976).

(15) T.M. BRODY, Action of Sodium Salicylate and related Compounds on tissue metabolism *in vitro*, J. Pharmac. Exp. Therap. 117, 39-51 (1956).

(16) M.W. Whitehouse, Uncoupling of Oxidative Phosphorylation in a Connective Tissue (Cartilage) and Liver Mitochondria by Salicylate Analogues: Relationship of Structure to Activity, Biochem. Pharmacol. 13, 319-336 (1964).

(17) R.W. GRADY, J.H. GRAZIANO, H.A. AKERS and A. CERAMI, The Development of New Iron-Chelating Drugs, J. Pharmac. Exp. Therap. 196, 478-485 (1976).

(18) D.G. MILLS, R.B. PHILP and M. HIRST, The Effects of Some Salicylate Analogues on Human Blood Platelets: 1. Structure Activity Relationships and the Inhibition of Platelet Aggregation, Life Sciences, 14, 659-772 (1974).

(19) M. ATIK and K. MATINI, Platelet Dysfunction: An Important Factor in Massive Bleeding from Stress Ulcer, The Journal of Trauma, 12, 834 (1972).

(20) G.V.R. BORN, Aggregation of Blood Platelets by Adenosine Diphosphate and its Reversal, Nature, 194, 927-929 (1962).

(21) J.F. MUSTARD, B. HEGARDT, H.L. ROWSELL and R.L. MacMILLAN, The Effect of Adenosine Nucleotides on Platelet Aggregation and Clotting Time, Journal of Laboratory and Clinical Medicine 64, 548-559 (1969).

(22) N.G. ARDLIE and P. HAN, Enzymatic Basis for Platelet Aggregation and Release: The Significance of the "Platelet Atmosphere" and the Relationship between platelet Function and Blood Coagulation", British J. of Haematology, 26, 331-356 (1974).

(23) R.J. LUKE and F. GIBSON, Location of Three Genes Concerned with Conversion of 2,3-Dihydroxybenzoate into Enterochelin in *Escherichia Coli* K-12, J. Bacteriol. 107, 557-562 (1971).

(24) F. GIBSON, Chorismic Acid. Biochemical Preparations, 12, 94-97 (1968).

(25) M. GLATT, B. PESKAR and K. BRUNE, Leukocytes and Prostaglandins in Acute Inflammation, Experientia, 30, 1257-1259 (1974).

(26) K. BRUNE, How Aspirin Might Work: A Pharmacokinetic Approach, Agents and Actions, 4, 230-232 (1974).

APPENDIX 1.

6-Methylsalicylic acid (AGN-816) and 6-methylaspirin (AGN-833) were kindly provided by the Nicholas Research Institute, Slough, England. Methyl 2-acetoxybenzoate was prepared from methyl salicylate (8) and recrystallised from petroleum ether (b.p.40-60), m.p. 51-2°. 2-acetoxybenzaldehyde, m.p. 38° (9), the salicylaldehyde dimer, dibenzo-2,6,9-bisdioxan, m.p. 129° and salixylaldehyde triacetate, m.p. 103° (10) were all prepared by treating salicylaldehyde with acetic anhydride. 2,3-dihydroxybenzoic acid (Fluka A.G., Buchs, Switzerland) was acetylated, using pyridine as a catalyst (11), yielding ·2,3-diacetoxybenzoic acid (3-acetoxyaspirin) with m.p. 157-164°. 2-hydroxyacetophenone (Fluka), acetylated by procedure (8), yielded the acetate, m.p. 90°.

Purity of all products was ascertained by TLC using silica gel plates (E. Merck, Darmstadt, D.B.R.) and the following solvent systems:
Benzene ether acetic acid-methanol (120:60:18:1, v/v)
Petroleum ether (b.p. 40-60°)-propionic acid (10:1, v/v)
Phenolic acetates gave no coloration with $FeCl_3$ in ethanol.

GASTROINTESTINAL AND OTHER SIDE-EFFECTS FROM THE USE OF ASPIRIN
AND RELATED DRUGS; BIOCHEMICAL STUDIES ON THE MECHANISMS OF
GASTROTOXICITY.

by K.D.Rainsford
Department of Biochemistry, University of Tasmania Medical
School, Hobart, Tasmania, Australia, 7001.

1. Main Side-Effects:

The main side-effects often associated with aspirin consumption
are gastrointestinal bleeding and ulceration, kidney damage,
congenital malformations, and hypersensitivity (1). Of these
gastrointestinal damage is the most important side-effect (1).
The evidence associating aspirin ingestion with the development
of renal nephropathy indicates that this may result from consumption of massive doses of the drug or from combinations with
other analgesic and/or anti-inflammatory drugs (e.g. phenacetin,
or phenylbutazone) (1,2,3). There are indications that, here in
Australia, dehydration may enhance the propensity of aspirin to
cause renal irritation (4); a feature which may be due to an
accumulation of high concentrations of the drug in tubules and
papillae of this organ. Changes in the renal concentrations of
the drug or its metabolites may also occur in individuals receiving drugs with diuretic actions (e.g. caffeine or alcohol)
so increasing the prospects of further damage.

In common with a variety of other drugs, aspirin and other salicylates may cause congenital malformations especially during
the first trimester of pregnancy (5,6). Previously, there were
indications, that the effect may be of low grade (6). Recent studies suggest that this problem could be a potentially frequent
occurrence than at first thought (at least in Australia) since
a significant number of women have been detected from urinalysis
as having consumed salicylates during pregnancy (7). As with the
renal damage this may prove more significant when combinations
of drugs are taken during the early stages of pregnancy. As a
partial explanation of the foetal toxic effects it seems that
the foetus could be especially sensitive to growth inhibitory
actions of the salicylates (8,9). Since the obvious way of preventing this side-effect is to avoid self-administration of the
drug during the first trimester of pregnancy (1) it would seem
appropriate to make the public more aware of this potential
side-effect.

Hypersensitivity, manifest in the form of skin rashes, asthma,
and nasal polyps, is an important side-effect, but the incidence
in the population is variable (1,10). This side-effect has been
discussed elsewhere along with side effects of minor importance
(1,10), (see also paper by McQueen in this Symposium).

2. Postulated Mechanism of Gastric Damage by Aspirin:

It is now clearly recognized that the mechanism of gastric

damage has a multifactorial basis (11). It is also useful to think of the development of the earlier stages of gastric damage in relation to the focal nature of the lesion which develops (Figure 1). In viewing the events which appear to de-develop, one of the first events envisaged is the sloughing of the protective mucus layer (12), which occurs by drug molecules aggregating the mucus components (13) (Figure 1). Drug particles then readily penetrate and aid destruction of the underlying surface cells and increase permeability of the mucosa (11). Discharge of mucus from the mucus globules in mucous cells occurs as a protective response of the mucosa to insult by aspirin (11). Where high concentrations of the drug (in solution or as particles) cause exhaustion of mucus from these cells, there is accompanying this sloughing of mucus. The underlying cells then become exposed to attack by the combined effects of the drug, acid and pepsin.

At the same time, high concentrations of aspirin accumulate in the acid-secreting parietal cells as a consequence of the unique pH gradient across the membranes exposed to the gastric pit (Figures 1 and 2) (14). Damage to these cells may be greatly enhanced by the fact that the acid-transporting canals (canaliculi) exist as finger-like extensions of parietal cells into the gastric pit. The resulting large surface area of this structure greatly facilitates absorption of drugs by these cells (15,16,17). Once the parietal cells have been disrupted they could act as a focus for damage to neighbouring cells and greatly weaken the structure of the mucosa (17). As with mucous cells damaged by aspirin, permeability changes may accompany physical effects on the membrane of the parietal cells (11). Also, it can be assumed that rupturing of the mucosal capillaries occurs; This is probably due to physical pressure from vasodilation and consequent fluid accumulation in tissues induced by the release of histamine (18), the suppression of prostaglandin release (19,20) or a block in the pressor response of noradrenaline (20). Bleeding from these capillaries may be enhanced by delayed bleeding time induced by aspirin. This effect is due largely to a reduction in platelet aggregation by the drug (21). It should be emphasized that the preceeding events are thought to occur within a short time (minutes) after administration of aspirin. Following this a variety of secondary biochemical effects and physiological responses ensue (for details see 11) which can be summarized:

> (a) A suppression of acid production occurs resulting from an inhibition of the production of ATP by uncoupling of oxidative phosphorylation and accelerated adenyl cyclase activity. A direct antagonism of acetylcholine actions could also be responsible for inhibiting acid production.
>
> (b) An inhibition of the production of the protective mucus layer occurs probably due as a consequence of the combined inhibitory effects of the drug on the

Figure 1.

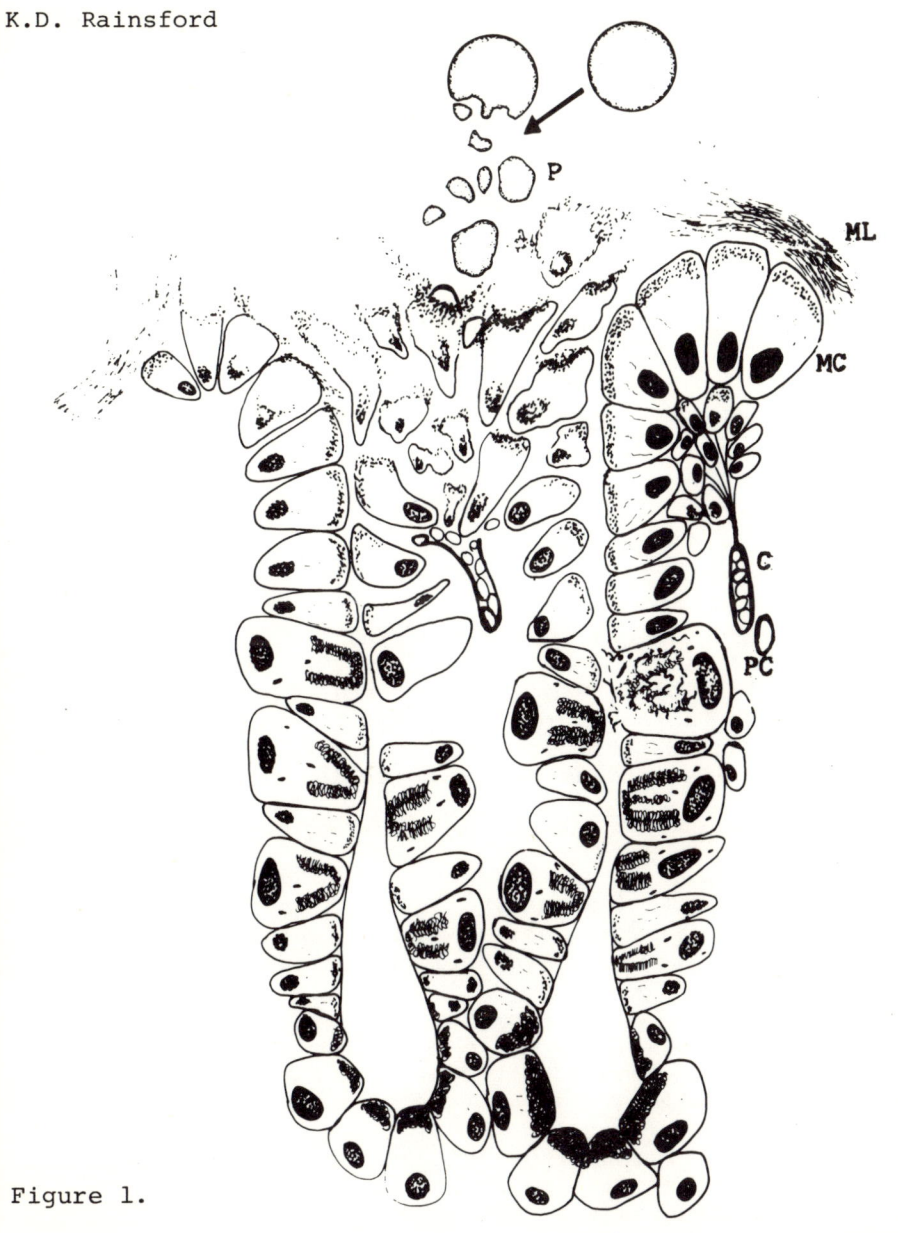

Diagrammatic representation of the events envisaged in the development of focal lesions induced by aspirin. When tablets or particles of the drug (P) come in contact with the surface mucosa they disperse and drug molecules cause aggregation and sloughing of the protective mucus layer (ML). Particles then penetrate this barrier and cause direct denaturation of underlying mucous-secreting cells (MC). Additional destruction of these neighbouring cells then occurs through the actions of acid and pepsin. At the same time there is an accumulation of large quantities of drug anions inside the acid-secreting parietal cell (PC) and this causes inhibition of the metabolic functions and disruption of the structure of the cell. Disruption of the capillaries (C) and prolongation of the bleeding time leads to focal haemorrhage. There follows infiltration of leucocytes and a variety of biochemical events affecting the production of mucus and acid in the stomach.

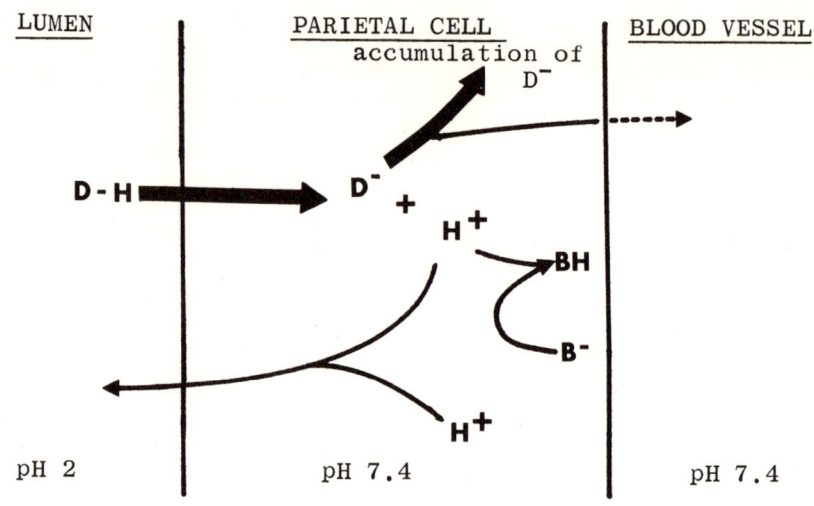

| LUMEN | PARIETAL CELL | BLOOD VESSEL |
| | accumulation of D⁻ | |

pH 2 | pH 7.4 | pH 7.4

Figure 2.

Accumulation of drug anions in the parietal cell. The pH gradient (pH 2 to 7.4) across the exposed parietal cell membranes favours rapid uptake of non-ionized drug molecules (15,16). Once inside these cells, dissociation occurs of protons from drug anions and these accumulate because of the slower rate of transit out of parietal cells into neighbouring cells or blood vessels.

        supply of ATP (necessary for the biosynthesis of mucus glycoproteins) as well as direct inhibition of enzymes involved in the biosynthetic reactions.

(c) Local tissue autolysis occurs from the release of lysosomal enzymes which could come from leucocytes accumulating at damaged sites as well as from damaged mucosal cells.

(d) A variety of inhibitory effects on the biosynthetic reactions in cells could be responsible for a delay in regeneration of the mucosa.

3. Aspirin and Ulceration:

Studies in laboratory animals indicate that regardless of time or level of oral dosage (within the therapeutic range) that aspirin alone does not produce ulcers per se (i.e. not in the strict pathological sense as characterized by perforation of the muscularis mucosa etc). Indeed it has been suggested that the gastric mucosa becomes refractory to damage by the drug after repeated oral administration (22). The logical question to ask at this stage is "if aspirin alone doesn't cause ulcer development, then what role does aspirin play in ulcer formation?"

ASPIRIN (50 mg/kg)

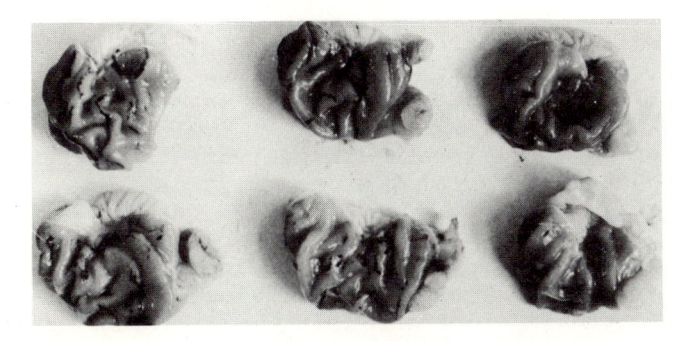

ASPIRIN (50 mg/kg) + STRESS

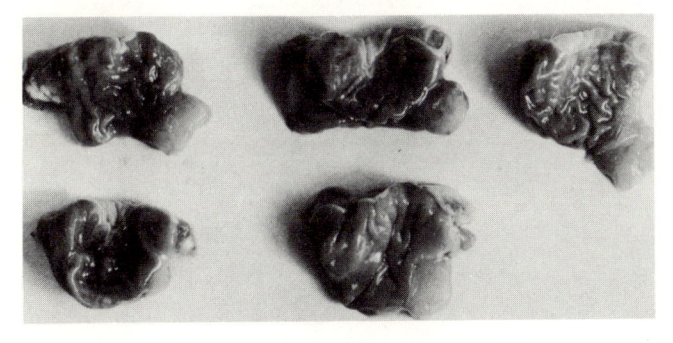

ASPIRIN (50 mg/kg) + STRESS + PGE$_1$

Figure 3.

The appearance of the gastric mucosa in starved (24 hrs) rats, 2 hrs after the administration of a low therapeutic dose of (50 mg/kg body weight) aspirin and simultaneous exposure to mild stress (cold, -15° for 30 mins). It can be seen that even under these mild dose/stress conditions that the number and severity of lesions in aspirin plus stress-treated animals increased dramatically. Concurrent administration of prostaglandin E$_1$ (2 mg/kg i.p.) completely eliminated the damage in aspirin plus stress treated animals. Other prostaglandins (PGE$_2$ or the long acting 15(S),15 methyl PGE) also diminished or abolished lesion development depending on the dose of aspirin employed. No lesions were observed in animals exposed to stress alone.

Several authors have suggested from clinical studies that other factors such as alcohol or stress may aid in the development of gastric damage initiated by aspirin (23,24) although the clinical evidence is equivocal (25). Following this suggestion it was found that stress (which mimicks anxiety or other types of psychological conditions) markedly sensitizes the stomach to the irritant actions of aspirin (Figure 3) (26). The effects were evident even at low doses of the drug and under mild stress conditions (Figure 3). In pigs, the repeated administration of aspirin and exposure to stress lead to the development of deep crater-like ulcers identical in appearance to those observed in humans (26). By contrast, no signs of damage were observed in control animals exposed to stress alone. In the animals given aspirin alone only focal lesions developed in the gastric mucosa (26). This and other studies (27) show that aspirin and other acidic non-steroid anti-inflammatory drugs interact synergistically with stress in the development of gastic damage (26,27).

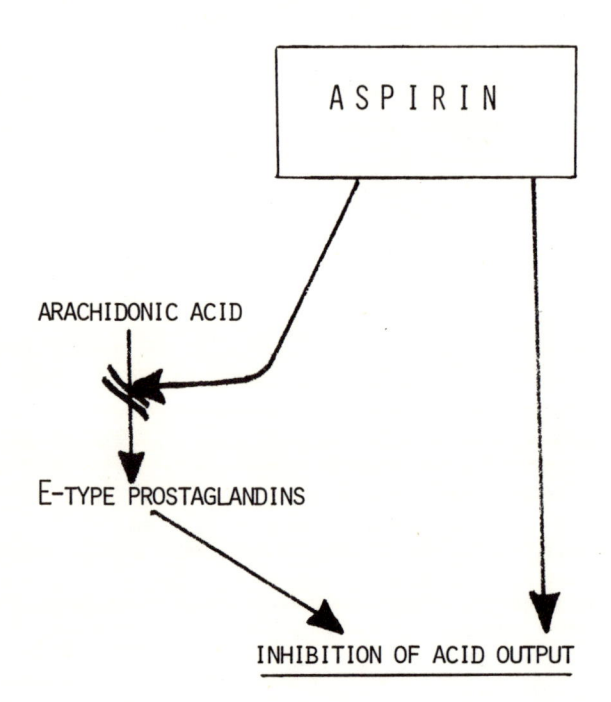

The synergy of physical stress (cold, restraint) plus aspirin has been demonstrated in rats, guinea pigs and pigs, and the effect in rats is manifest to the same degree in both females and males (26,27,28). It is very specific, in that damage is only confined to the acid secreting regions of the gastric mucosa in these species. There was no indication of any irritant effects in the remainder of the intestinal tract, and this was also the case for all acidic N.S.A.I. drugs studied (26,27). By contrast, no damage was evident to animals given known non-ulcerogenic analgesics (such as paracetamol or dextropropoxyphene) (26). Recent studies (27,30) show that "disease stress" conditions (e.g. acute inflammation induced by carrageenan or adjuvant arthritis) do not enhance the irritant effects of aspirin in the gastrointestinal tract. By comparison, indomethacin has been found to produce more intestinal damage in adjuvant-treated as compared to control rats (31). These observations further demonstrate the specificity of aspirin interactions with stress conditions.

One of the main effects involved in the development of the aspirin plus stress induced damage in the stomach seems to be attributable to the vagal stimulation of acid release and accompanying local histamine release (28,29). The E-type prostaglandins which alledgedly possess anti(acid)-secretory actions definitely block the damage due to the combination of aspirin and physical stress (Figure 3) (27,28). It is tempting to suggest that the inhibition of prostaglandin release or production may abet or cause the release of acid. Before we accept this proposition it is appropriate to consider the action of aspirin and the prostaglandins alone on the acid secreting capacity of the gastric mucosa. The E-type prostaglandins as well as aspirin and its metabolite salicylate inhibit acid output (32,33,34,35). Also, the antisecretory actions of the prostaglandins have been associated with the inhibitory effects of these agents on ulcer formation under a variety of experimental conditions (32,36). The E-type prostaglandins also prevent damage induced by aspirin alone (11). Thus we have a paradoxical situation (summarized in Figure 4) where aspirin and the E-type prostaglandins both inhibit acid secretion. For aspirin to have inhibitory effects on prostaglandin production and therefore to allow acid secretion to increase which has suggested (37) would seem untenable. When we examined the influence of prostaglandin E_2 on the output of acid and other physiological ions in aspirin plus stress-treated rats the situation seemed even more complex as the data in Table 1 shows. The content of H^+ ions in the lumen of aspirin plus stress treated rats was actually reduced compared with that in animals given aspirin alone. However, administration of PGE_2 did not decrease the acid content in the lumen of aspirin plus stress treated animals but restored the values to those obtained from administration of aspirin alone. There was no effect on the permeability of the mucosa to other ions (Na^+, K^+, Cl^-). So no generalized permeability changes can be envisaged in aspirin plus stress-treated rats given prostaglandins. To look for an answer to these observations it is probably necessary to consider the possibility of effects of aspirin on blood flow and tissue vascularization (20,38). Lindt and Baggiolini (39) have recently shown that phenylbutazone and stress both decrease the blood content in various regions of the gastric mucosa of rats. The blood content was markedly increased following administration of a prostaglandin E_2 analogue (38). A similar situation may occur in that aspirin may cause a reduction in blood content and tissue vascularization so promoting an ischaemic condition in the stomach. The combination of stress with aspirin could have an added effect in reducing blood flow.

From these studies it is clear that in assaying gastric irritancy of anti-inflammatory compounds the aspect of stress should be considered since this is of major clinical significance. Also, the physical stress model affords a means by which rapid and sensitive assay can be obtained of any potential irritant actions of N.S.A.I. drugs.

4. Towards an Aspirin without Gastric Irritant Side-effects:

Possible strategies for reducing the ulcerogenic actions of aspirin by enhancing protective responses of the mucosa or neutralizing the gastric contents have been discussed elsewhere in this symposium (see Whitehouse, these Proceedings). As an approach towards reducing the ulcerogenity of aspirin we have recently commenced a study of the structure-activity relationships involved in the gastric irritation produced by the salicylates (30). The aim being to establish the features which make this group of drugs irritant and to see if there are ways of reducing this side-effect without affecting the anti-inflammatory/analgesic potency of the drugs. Unfortunately, attenuation of the gastric irritant effects of aspirin often reduces the efficacy of these derivatives (30). Several commercially available salicylate derivatives showed less gastric irritancy than aspirin including salicyl salicylic acid, benorylate(R) (the paracetamol ester of aspirin) and flufenisal(R) (5-(4'-fluorophenyl)acetyl salicylic acid) but in the case of benorylate this was accompanied by some reduction in efficacy (30). It seems that some fundamental consideration of the structure-activity relationships is therefore warranted.

TABLE 1

Luminal Ion Content After Aspirin and/or Stress Treatments in Rats. (Content in umoles (mean ± S.O.).

Treatment	H^+	Na^+	K^+	Cl^-
Control	8.4±4.0	206±31	173±127	393±141
Control + Stress	4.5±0.9[d]	117±115	118±96	117±76
Aspirin	53.2±19.3[e]	778±483[e]	505±463	526±285
Aspirin + Stress	19.2±8.7[ac]	1252±572[e]	572±464[c]	293±426[c]
Aspirin+Stress+PGE_2	45.7±16.2[b]	1070±449	584±493	598±268

Significant difference ($P \leq 0.05$) between aspirin and aspirin + stress groups (a); between aspirin + stress and aspirin + stress + PGE_2 groups (b); between control + stress and aspirin + stress (c); between control and control + stress (d), control + aspirin (e). 4 animals per group.
H^+ content determined by titration to pH 7 with 0.01M NaOH, Na^+ and K^+ content determined by flame photometry and Cl content by chloridimetry all at 60 mins. after dosing with drugs.

In a brief review of screening studies performed on salicylate analogues at Nicholas Research Institute Laboratories, Anderson stated that gastric irritancy in guinea pigs depends on the presence of an unsubstituted carboxyl group and given good absorption, the ortho substituent of the benzoic acid derivative also influences irritation to the mucosa (40). In essence we have confirmed this observation in rats (30) and examined the role of the acetyl group in lesion development. Although the presence of the acetyl group does enhance gastric irritancy of salicylic acid and closely related derivatives, it is clear that this does not apply to all salicylate derivatives. The modifying effects of other ring substituents or the elimination of the acidity of the carboxylic acid moiety of the salicylates does reduce gastric irritancy (30). Reduction of the acidic characteristic such as by formation of esters of the parent acid i.e. acetylsalicylic acid, was found to markedly reduce irritation without apparently affecting therapeutic i.e. anti-inflammatory activity in normal and stressed (cold, disease) rats (30). Moreover, we have recently extended these studies to observations in pigs; a species with gastrointestinal structure and functions and dietary status (i.e. omnivorous) more closely related to that of man than other laboratory mammals. It appears, therefore that pigs may be a more suitable species to study gastric irritation and ulcer development from N.S.A.I. drugs as previous work has indicated (26). In conclusion, it appears that modification of the acidic characteristics of the carboxyl group is one satisfactory way to modify the irritant actions of aspirin while still retaining full anti-inflammatory potency.

Summary:

A multifactorial basis has been shown to exist in the development of gastric damage induced by aspirin and related N.S.A.I. drugs. Aspirin-induced gastric damage is characterized by a variety of physical and biochemical changes induced in the gastric mucosa which occur at different stages after administration of the dug.

Aspirin only causes gastric ulceration and massive haemorrhage in the stomach when the stomach has been sensitized by the prior exposure to moderate stress conditions (which may resemble anxiety or psychologic stress). A model of ulcer development in which aspirin or other N.S.A.I. drugs are given to rats or pigs exposed to brief periods of stress has been described. Using this more sensitive assay procedures can be explored for reducing the gastric damaging effects of N.S.A.I. drugs. One such procedure involves chemical modification of the carboxylic acid moiety of aspirin.

Acknowledgements:

These studies were funded by grants from the National Health and Medical Research Council (Australia) and the University of Tasmania Research Committee.

REFERENCES

(1) K.D.RAINSFORD, Aspirin; Actions and Uses, Aust.J.Pharm. 56, 373-382 (1975).

(2) NEW ZEALAND RHEUMATISM ASSOCIATION STUDY, Aspirin and the Kidney, Brit.med.J. 1, 593-596 (1974).

(3) A.F.MACKLON, A.W.CRAFT, M.THOMPSON and D.N.S.KERR, Aspirin and Analgesic Nephropathy, Brit.med.J. 1, 597-600 (1974).

(4) R.S.NANRA and P.KINCAID-SMITH, Chronic Effects of Analgesics on the Kidney, Progress in Biochemical Pharmacology 7, 285-323 (1972).

(5) I.D.G.RICHARDS, Congenital Malformations and Environmental Influences in Pregnancy, Brit.J.Prevent.Soc.Med. 23, 218-225 (1969).

(6) M.M.NELSON and J.O.FORFAR, Associations between Drugs Administered during Pregnancy and Congenital Malformations of the Foetus, Brit.med.J. 1, 523-527 (1971).

(7) G.TURNER and E.COLLINS, Foetal Effects of Regular Salicylate Ingestion in Pregnancy, Lancet 2, 338-339 (1975).

(8) K.S.LARSSON and H.BOSTROM, Teratogenic Action of Salicylates Related to the Inhibition of Mucopolysaccharide Synthesis, Acta Pediat.Scand. 54, 43-48 (1965).

(9) K.JANAKIDEVI and M.J.H.SMITH, Effects of Salicylate on the Incorporation of Orotic Acid into Nuclei Acids in vivo, J. Pharm.Pharmac. 22, 51-55 (1970).

(10) M.J.H.SMITH, Toxicology, in: The Salicylates : A Critical Bibliographic Review (Eds. M.J.H.Smith and P.K.Smith, Wiley-Interscience, New York) (1966) pp. 233-306.

(11) K.D.RAINSFORD, The Biochemical Pathology of Aspirin-induced Gastric Damage, Agents and Actions 5, 326-344 (1975).

(12) J.L.A.ROTH and A.VALDES-DAPENA, Topical Actions of Salicylates on the Buccal Mucosa in Man and on the Stomach in the Cat in: Salicylates: An International Symposium (Eds. A.St. J. Dixon, B.K.Martin,M.J.H.Smith and P.H.N. Wood, Churchill, London, 1963) pp. 224-225.

(13) K.D.RAINSFORD, J.WATKINS and M.J.H.SMITH, Aspirin and Mucus J.Pharm.Pharmac. 20, 941-943 (1968).

(14) K.D.RAINSFORD and K.BRUNE, Role of the Parietal Cell in Gastric Damage Induced by Aspirin and Related Drugs: Implications for Safer Therapy, Med.J.Aust., in press (1976).

(15) L.S.SCHANKER,P.A.SHORE,B.B.BRODIE and C.A.M.HOGBEN, Absorption of Drugs from the Stomach. 1. The Rat, J.Pharmac.Exp. Therap. 120, 528-539 (1957).

(16) B.K.MARTIN, Accumulation of Drug Anions in Gastric Mucosal Cells, Nature (Lond) 198, 896 (1963).

(17) K.D.RAINSFORD, Electronmicroscopic Observations on the Effects of Orally Administered Aspirin and Aspirin-Bicarbonate Mixtures on the Development of Gastric Mucosal Damage in the Rat, Gut 16, 514-527 (1975).

(18) L.R.JOHNSON, Histamine Liberation by Gastric Mucosa of Pyloric Ligated Rats Damaged by Acetic and Salicylic Acids, Proc.Soc.Exp.Biol. (N.Y.) 121, 384-386 (1966).

(19) R.FLOWER, R.GRYGLEWSKI, K.HERBACZYNSKA-CEDRO and J.R.VANE, Effects of Anti-inflammatory Drugs on Prostaglandin Biosynthesis, Nature (New Biol.) 238, 104-106 (1972).

(20) D.F.HORRIBIN, M.S.MANKU, R.KARMALI, B.S.NASSAR and P.A. DAVIES, Aspirin, Indomethacin, Catecholamines and Prostaglandin Interactions on Rat Arterioles and Rabbit Hearts, Nature (Lond) 250, 425-426 (1974).

(21) J.WEISS and L.M.ALEDORT, Impaired Platelet/Connective Tissue Reaction in Man after Aspirin Ingestion, Lancet 2, 495-497 (1968).

(22) J.W.HURLEY and L.A.CRANDALL, The Effect of Various Salicylates on the Dog's Stomach: A Gastroscopic Photographic Evaluation, in: Salicylates. An International Symposium. (Eds. A.St.J.Dixon, B.K.Martin,M.J.H.Smith and P.H.N.Wood, Churchill, London, 1963) pp. 213-216.

(23) M.I.GROSSMAN, K.K.MATSUMOTO and R.S.LICHTER, Faecal Blood Loss Produced by Oral and Intravenous Administration of Various Salicylates, Gastroenterology 40, 383-388 (1961).

(24) G.H.JENNINGS, Causal Influences in Haematemesis and Melaena, Gut 6, 1-13 (1965).

(25) M.J.S.LANGMAN, Epidemiological Evidence for the Association of Aspirin and Acute Gastrointestinal Bleeding, Gut 11, 627-634 (1970).

(26) K.D.RAINSFORD, A Synergistic Interaction between Aspirin, or other Non-steroidal anti-inflammatory Drugs, and Stress which Produces Severe Gastric Mucosal Damage in Rats and Pigs, Agents and Actions 5, 553-558 (1975).

(27) K.D.RAINSFORD, The Effects of Physical Stress vis à vis Disease, and other Forms of Stress on the Development of Aspirin-induced Gastric Damage, in preparation (1976).

(28) K.D.RAINSFORD, Some Factors Involved in the Synergistic Interaction between Aspirin and Stress which Results in Severe Gastric Mucosal Damage, Clin.Exp.Physiol.Pharmacol., in press (1976).

(29) P.A.BROWN, T.H.BROWN and J.VERNIKOS-DANELLIS, Histamine H_2 Receptor: Involvement in Gastric Ulceration, Life Sci. 18, 339-344 (1976).

(30) K.D.RAINSFORD and M.W.WHITEHOUSE, Gastric Irritancy of Aspirin and its Analogues: Anti-inflammatory Effects without this Side-effect, J.Pharm.Pharmac., in press (1976).

(31) G. DI PASQUALE and P.WELAJ, Ulcerogenic Potential of Indomethacin in Arthritic and Non-Arthritic Rats, J.Pharm. Pharmac. 25, 831-832 (1973).

(32) A.ROBERT, J.E.NEZAMIS and J.P.PHILLIPS, Effect of Prostaglandin E_1 on Gastric Secretion and Ulcer Formation in the Rat. Gastroenterology 55, 481-487 (1968).

(33) M.S.AMER and G.R.MCKINNEY, Cyclic AMP and the Mechanism of Action of Gastrointesinal Hormones, in: Chemistry and Biology of Peptides, Proceedings, (Ed. J.Meinhofer, Ann Arbor Science, Ann Arbor, 1972) pp. 617-620.

(34) H.W.DAVENPORT, Gastric Mucosal Haemorrhage in Dogs. Effect of Acid, Aspirin and Alcohol, Gastroenterology 56, 439-449 (1969).

(35) T.GLARBORG-JORGENSEN, E.L.KAPLAN and G.W.PESKIN, Salicylate Effects on Gastric Acid Secretion, Scand.J.clin.Lab.Invest. 33, 31-38 (1974).

(36) Y.H.LEE, W.D.CHENG, R.G.BIANCHI, K. MELLISON and J.HANSEN, Effects of Oral Administration of PGE_2 on Gastric Secretion and Experimental Peptic Ulcerations, Prostaglandins 3, 29-45 (1973).

(37) S.H.FERREIRA and S.R.VANE, New Aspects of the Mode of Action of Non-Steroid Antiinflammatory Drugs, Ann.Rev.Pharmac. 14, 57-73 (1974).

(38) A.BENNETT, I.F.STAMFORD and W.G.UNGAR, Prostaglandin E_2 and Gastric Acid Secretion in Man, J.Physiol. (Lond) 229, 349-360 (1973).

(39) S.LINDT and M.BAGGIOLINI, Effect of a $PG-E_2$ Analogue ($PG-E_2'$) on the Vascularization of the Gastric Mucosa in the Rat, Experientia 32, 802 (1976).

(40) K.W.ANDERSON, Some Biochemical and Physiological Aspects of Salicylate-Induced Gastric Lesions in Laboratory Animals, in: Salicylates: An International Symposium (Eds. A.St.J. Dixon, B.K.Martin, M.J.H.Smith and P.H.N.Wood, Churchill, London, 1963) pp. 217-223.

GASTRIC ULCER, ASPIRIN ESTERASE AND ASPIRIN

by K.D.Landecker, J.E.Wellington, J.H.Thomas and D.W.Piper
Royal North Shore Hospital of Sydney, Sydney, N.S.W. Australia

The role of aspirin ingestion in the cause and persistence of gastric ulcer disease is still not clear. This disease is the result of interaction of many factors, most of which are still in dispute. Acid and pepsin are regarded as indispendsable factors (1) which act against the normal resistance of the gastric mucosa to self-digestion. Other factors relevant to chronic ulcer disease may be genetic, psychosomatic, continuance of smoking and the consumption of alcohol and aspirin. Bile reflux and delayed gastric emptying have also been incriminated and in some patients the presence of gastritis (2).

Acute ulcers which heal without scarring, but manifest their presence by haemorrhage and even perforation. These can be considered as a separate entity, as they can be caused by severe physical stress, such as trauma, infection or anoxia, as well as by drugs such as aspirin, phenylbutazone and indomethacin.

Aspirin, either by direct absorption or via the circulation, enters the cell. Here it is postulated to interfere with the cell's energy metabolism, resulting in decreased mucus synthesis and increased cell loss by exfoliation. Thus the mucosal barrier is broken, resulting in increased back-diffusion of hydrogen ions into the cell layer (4). The cells are swollen and damaged at the intercellular junctions by their increased uptake of hydrogen ions (3). The cells may then shed at an increased rate and histamine released with further secretion of acid and pepsin so leading to damage to submucosal capillaries with necrosis and bleeding (4). Added to this is the known factor of impaired platelet aggregation caused by aspirin (6). Bile has a detergent action on the mucus and epithelial cells, allowing increased back-diffusion of hydrogen ions also. In a study in which the changes in electrical potential difference across the mucosa were measured as a marker of disruption of the mucosal barrier (7) taurocholic acid was found to increase significantly the mucosal damage caused by aspirin.

It is certainly true that acetylsalicylic acid ingestion causes increased gastric blood loss. This has been measured by studies using radio-active chromium labelled red cells. It has been shown that the loss of blood from aspirin ingestion is greater than that caused by sodium salicylate (8). This loss is increased by simultaneous ingestion of alcohol whereas there is no loss of blood produced by alcohol alone (9).

Since among patients with chronic gastric ulcer, those who continue to take aspirin more commonly have large ulcers than those who do not (10), further evidence of an association was sought. Patients with chronic gastric ulcer were matched at their initial diagnosis with controls selected by a series of random numbers from the electoral roll. These controls were then matched for age, sex and social grade, with the patients.

All patients and controls were asked what medicines they had been taking recently, and a specimen of urine was also collected for analysis for urinary salicylates.

In this method, to be published (11), the salicylates in the urine are first extracted into chloroform and then detected qualitatively by a spectrofluorometric method. The excitation wavelength is set at 305 nm and the fluorescence emission spectrum between 320 and 500 nm is recorded. An emission peak at 450 nm indicates the presence of salicylates (Figure 1). When the concentration of salicylate in the urine sample is very low (less than 5 mg/ml of urine) the emission peak at 450 nm is sometimes difficult to detect, owing to an interfering peak at 375 nm. This problem is overcome by drawing a line of best fit through the curve between 410 nm and 425 nm. If this line is found to cut the zero baseline after the 500 nm mark, the sample is considered to contain salicylate. The sensitivity of the method compared with "Phenistix" testing can be seen in Table 1. It is seen that as little as one Disprin tablet (Reckett and Colman; containing 300 mg soluble calcium aspirin) taken 24 hours previously still gives a positive result by this method. (This dose is not detected by Phenistix after 18 hours). Sodium salicylate dissolved in distilled water, was added to a urine specimen, previously analysed as negative, in various concentrations. Aliquots of 1 ml of each prepared solution were tested for salicylates by the previously described fluorometric method. The remainder was tested by "Phenistix" by three different observers. The order of concentrations was randomised for all tests to avoid bias.

EMISSION SPECTRUM OF SALICYLATES IN URINE

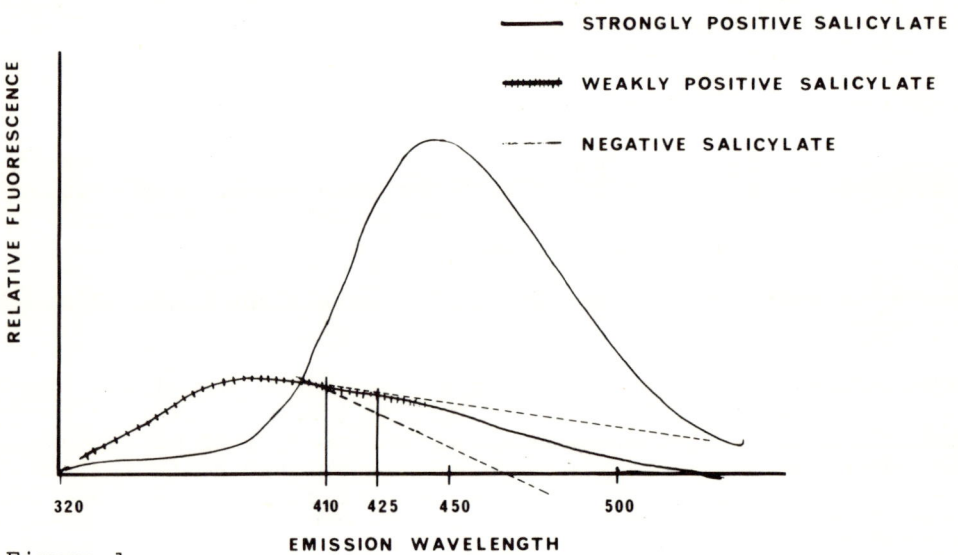

Figure 1

TABLE 1

Comparison of Phenistix and Fluorometric Methods for Detection of Urinary Salicylates

Concentration of sodium salicylate in urine (ug/ml)	Fluorometric test for salicylates	Phenistix Observer 1	Phenistix Observer 2	Phenistix Observer 3
500 ug/ml	+	+	+	+
250 ug/ml	+	+	+	+
100 ug/ml	+	+	?	+
50 ug/ml	+	+	−	?
20 ug/ml	+	−	−	?
10 ug/ml	+	−	−	−
5 ug/ml	+ (trace)	−	−	−
2.5 ug/ml (Approx. equivalent to result of one disprin after 24 hours)	+ (trace)	−	−	−
1.25 ug/ml	−	−	−	−

TABLE 2

Incidence of Salicylates in Urine of Chronic Gastric Ulcer Patients and Controls

Evidence of Salicylates	Gastric Ulcer Patients (42)	Controls Matched for Age, Sex and Social Grade from General Population (42)	Significance
History of Ingestion	10 (23.8 %)	4 (9.5 %)	0.1 > P > 0.05
Positive Result by Spectrofluorometry	20 (47.6 %)	12 (28.5 %)	0.1 > P > 0.05

TABLE 3

Evidence of Salicylates in Urine of Patients with Acute Upper Gastrointestinal Haemorrhage and Controls

Endoscopic Diagnosis	Patients with Positive Urinary Salicylates	Matched Controls with Positive Urinary Salicylates	Significance
Acute Ulcer G.U. or D.U. 11	5	3	P > 0.4
Chronic Ulcer G.U. or D.U. 7	1	2	
Total Ulcer Bleeds 18	6	5	P > 0.7
Bleeding from Other Causes 4	0	1	
Total 22	6 (27 %)	6 (27 %)	

Using this method, it was found that the group of 42 patients with chronic gastric ulcer had salicylates present in their urine almost twice as commonly (20 of the 42 patients) as did people in the control group (12 of the 42 controls) (Table 2). It is also seen from this table that historical evidence of salicylate ingestion is completely unreliable, compared with objective measurement, in both patient and control groups. The figure of 9.5 % in the control group, of people stating that they had taken salicylate containing preparations recently, does however conform with statements from surveys of analgesic consumption in the general population. The historical findings agree also with other published work on patients with chronic gastric ulcer. For example, Cameron (12) found a singificant increase in heavy aspirin intake (defined as 15 or more aspirin tablets each week) in patients with chronic gastric ulcer, compared to age and sex-matched controls (in this case patients with gallostones, hiatus hernia or colonic polyps). His patients answered a questionnaire on their first referral to the gastro-enterology clinic, before being interviewed by the physician. When patients who took aspirin for relief of alimentary symptoms and those whose symptoms preceeded aspirin use were excluded, the difference remained significant, between aspirin consumption in the patient and control groups.

The fluorometric method was also used to detect salicylates in a group of patients admitted to hospital with acute upper gastrointestinal haemorrhage. For each patient a control was selected, namely the emergency admission for some other disease, nearest in time and matched for age and sex. The urine was collected as soon as possible after admission, and tested by the method described. In this study (which is still in progress) no significant differences have been found to date between groups of patients with bleeding from various causes, and their matched controls (Table 3). Diagnosis was made by early endoscopy. These results do conflict with many previous findings, based on historical evidence (for example, see Levy (13)). However, Langman (14) has suggested that it is not possible to state that aspirin plays a major role in inducing gastro-duodenal bleeding, because of the weaknesses inherent in these controlled retrospective studies based on historical evidence alone.

Because acetylsalicylic acid causes more damage to the stomach than salicylic acid, the enzyme aspirin esterase which converts acetylsalicylic acid to salicylic acid in the body, was studied. Menguy and co-workers (15) have measured this enzyme group in plasma for normal males and females, excluding known heavy aspirin users, and have found significantly lower levels in females. From this they postulated a mechanism for the findings in Australia, that of an increased incidence of chronic gastric ulcer in middle-aged females who were heavy aspirin users (16, 17, 18, 19, 20). They suggested that these people were unable to metabolise aspirin as quickly, leaving the more toxic acetylsalicylic acid in contract with the gastric mucosa for longer periods.

pH ACTIVITY CURVE

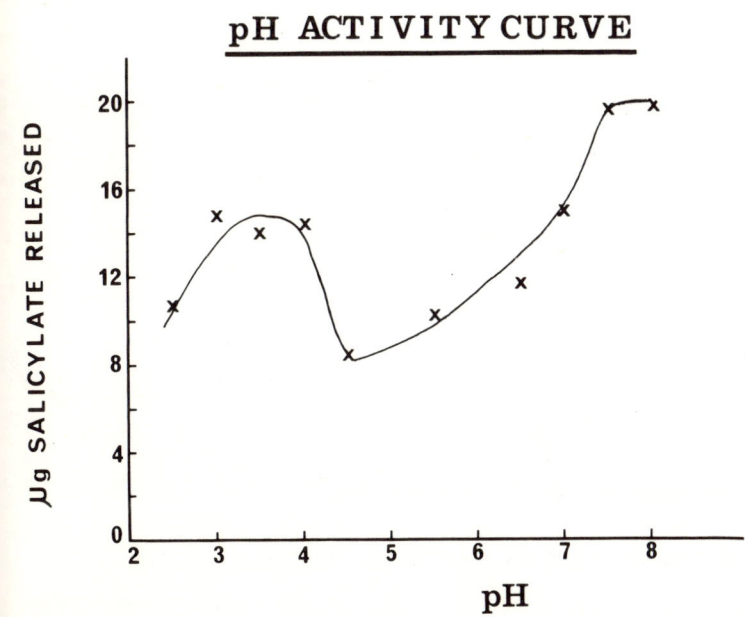

Figure 2

In the present study, the gastric mucosal aspirin esterase activity was measured in a group of patients who had gastric surgery for either gastric or duodenal ulcer, gastric carcinoma, or a number of other diseases (the latter being used as a control group). An homogenate of the mucosa was incubated with acetylsalicylic acid at pH 7.3 in buffered solution. The salicylate released was measured spectrofluorometrically and results expressed as ug of salicylate released per mg of protein in the mucosal homogenate. A pH activity curve (Figure 2) shows two peaks at which the enzyme appears to be most active, hence the probability that the enzyme in fact is a group of at least two isoenzymes. This deserves further study; however, it is seen from Table 4 that age, sex, origin of tissue (whether from fundus or antrum of stomach) have shown no significant differences between the results obtained. In these same patients the plasma aspirin esterase was also measured and no significant differences were found, although some age differences were noted (Table 5). The greatest difference was found between duodenal ulcer and gastric carcinoma groups. The significance of this is not as yet clear.

Since alcohol enhances the effect of aspirin on the stomach, experiments were done to study the effects of alcohol on the enzyme. The enzyme was increasingly inhibited as the alcohol concentration rose (Figure 3); alcohol, however, increased the spontaneous breakdown of acetylsalicylic acid. The net effect is a slowing in the conversion of acetylsalicylic acid to salicylic acid (Figure 4). An experiment using beer and whisky showed inhibition consistent with the alcohol content of these beverages. This result may explain the additive effects of alcohol and aspirin in the causation of gastric blood loss (9).

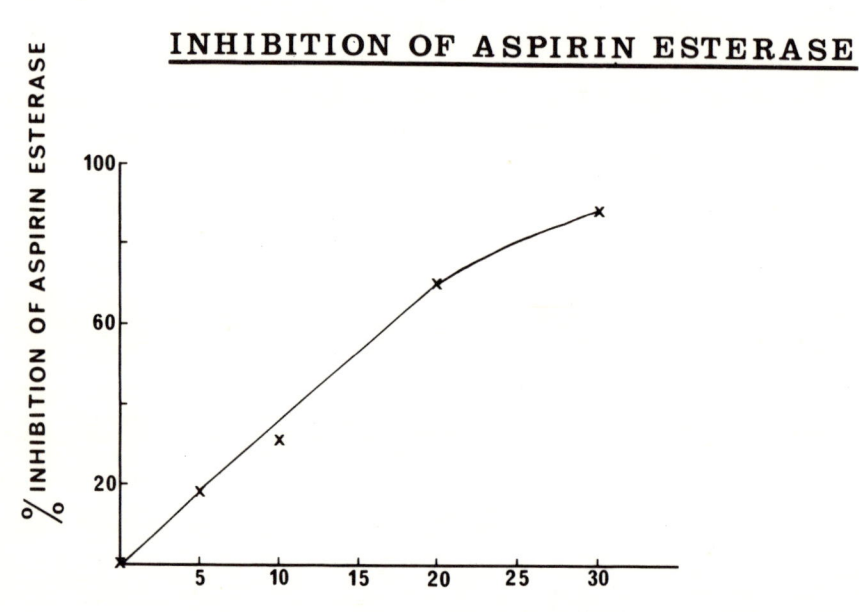

INHIBITION OF ASPIRIN ESTERASE

Figure 3

TABLE 4

Mucosal Aspirin Esterase Activity

Category	Group	(n)	Value	Significance
Age	55 years	(23)	6.5 ± 0.42	P = 0.3
	55 years	(21)	7.2 ± 0.41	
Sex	Male	(29)	6.6 ± 0.36	P = 0.4
	Female	(15)	7.2 ± 0.53	
Tissue	Body	(7)	7.9 ± 0.62	P = 0.4
	Antral	(7)	8.4 ± 0.56	
Disease	G.U.	(14)	7.5 ± 0.54	Greatest
	D.U.	(8)	5.9 ± 0.68	Difference
	Ca	(13)	6.9 ± 0.55	D.U. v. G.U.
	Control	(6)	6.9 ± 0.84	P = 0.1

TABLE 5

Plasma Aspirin Esterase Activity

Category	Group	(n)	Value	Significance
Age	55 years	(21)	59.5 ± 3.7	P = 0.01
	55 years	(14)	45.5 ± 3.3	
Sex	Male	(24)	52.3 ± 3.7	P = 0.3
	Female	(11)	57.4 ± 3.7	
Age + Sex	Male			
	55 years	(15)	59.4 ± 4.6	P = 0.01
	55 years	(9)	40.5 ± 3.8	
	Female			
	55 years	(6)	59.6 ± 3.8	P = 0.5
	55 years	(5)	54.7 ± 3.7	
Disease	G.U.	(10)	52.9 ± 4.0	Only
	D.U.	(8)	65.9 ± 6.7	Difference
	Ca	(5)	46.0 ± 4.6	Du Cancer
	Control	(10)	51.9 ± 5.7	P < 0.05

At present, the patients with chronic gastric ulcer are being studied again in remission, that is, when they are free of pain, to see whether they are then also taking more aspirin than the control group. It would also be interesting to know (but would seem impossible to solve) whether or not the patients in recurrence are taking aspirin prior to the onset of pain, or whether pain encourages them to take more aspirin.

EFFECT OF ALCOHOL ON THE ENZYMIC & NON-ENZYMIC HYDROLYSIS OF ASPIRIN

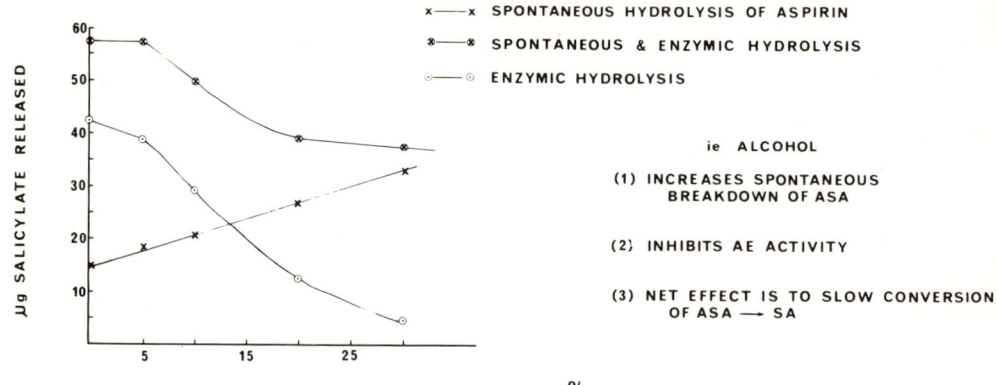

Figure 4

In conclusion, it is seen that a sensitive spectrofluorometric method is of great value in detecting urinary salicylates in order to provide objective evidence of salicylate ingestion. Patients must always be compared with an adequate control group. Aspirin esterase may have no role in aspirin induced gastric ulcer disease, but should be studied further as regards factors causing its activation or inhibition and the possibility of its being in fact a group of isoenzymes.

REFERENCES

(1) K.S.IVEY, Review - Pathogenesis of peptic ulcer, Aust.N.Z. Med.J. 4, 71-74 (1974).

(2) B.RHODES and J.CALCRAFT, Aetiology of Gastric Ulcer in Clinics, in: Gastroenterology - Peptic Ulcer, Ed.Sircus, Saunders Publishing Co. (1973).

(3) L.DESBAILLETS, Gastric Mucosal Injury by Anti-inflammatory drugs, in: Advances in Clinical Pharmacology 6 (1974).

(4) H.W.DAVENPORT, Damage to the Gastric Mucosa: Effect of Salicylates and Stimulation, Gastroenterology, 49, 189-196 (1965).

(5) B.FRENNING and K.J.OBRINK, The Effects of Acetic and Acetylsalicylic acids on the Appearance of the Gastric Mucosal Surface Epithelium in the Scanning Electron Microscope, Scand. J.Gastroent. 6, 605-612 (1971).

(6) H.J.WEISS, L.M.ALEDORT and S.KOCHWA, The Effect of Salicylates on the Haemostatic Properties of Platelets in Man, J.Clin.Invest. 47, 2169-2180 (1968).

(7) K.M.COCHRAN, J.F.MACKENZIE and R.I.RUSSELL, Role of Taurocholic Acid in the Production of Gastric Muscosal Damage after Ingestion of Aspirin, Brit.Med.J. 1, 183-185 (1975).

(8) J.R.LEONARDS and G.LEVY, Gastrointestinal Blood Loss from Aspirin and Sodium Salicylate Tablets in Man, Clin.Pharm. Therap. 14, 62-66 (1973).

(9) K.GOULSTON and A.R.COOKE, Alcohol, Aspirin and Gastrointestinal Bleeding, Brit.Med.J. 4, 664-665 (1968).

(10) R.P.HERRMAN and D.W.PIPER, Factors Influencing the Healing Rate of Chronic Gastric Ulcer, Am.J.Dig.Dis. 18, 1-6 (1973).

(11) J.WELLINGTON, K.D.LANDECKER, J.H.THOMAS and D.W.PIPER, Spectrofluorometric Detection of Urinary Salicylates, in preparation (1976).

(12) A.J.CAMERON, Aspirin and Gastric Ulcer, Mayo Clin.Proc. 50, 565-570 (1975).

(13) M.LEVY, Aspirin Use in Patients with Major Upper Gastrointestinal Bleeding and Peptic Ulcer Disease, New Engl.J.Med. 290, 1158-1162 (1974).

(14) M.J.S.LANGMAN, Controversy in Internal Medicine II, Ed.F.J. Ingelfinger, Saunders Publishing Co. (1974).

(15) R.MENGUY, L.DESBAILLETS and Y.F.MASTRES, Evidence of a Sex Linked Difference in Aspirin Metabolism, Nature (Lond) 239, 102-103 (1972).

(16) B.P.BILLINGTON, The Australian Gastric Ulcer Change: Intestate Variations, Aust.Ann.Med. 12, 153-159 (1963).

(17) B.P.BILLINGTON, Observations from New South Wales on the Changing Incidence of Gastric Ulcer in Australia, Gut 6, 121-133 (1965).

(18) B.L.CHAPMAN and J.M.DUGGAN, Aspirin and Uncomplicated Peptic Ulcer, Gut 10, 443-450 (1969).

(19) J.M.DUGGAN and B.L.CHAPMAN, The Incidence of Aspirin Ingestion in Patients with Peptic Ulcer, Med.J.Aust. 1, 797-800 (1970).

(20) M.D.GILLIES and A.SKYRING, Gastric Ulcer and Duodenal Ulcer; The Association Between Aspirin Ingestion, Smoking and Family History of Ulcer, Med.J.Aust. 2, 280-285 (1969).

ASPIRIN AND ULCERS.

by J.M.Duggan
Royal Newcastle Hospital, Newcastle, N.S.W., Australia

In this paper the evidence linking chronic gastric ulcer and regular aspirin use will be reviewed. Fifteen years ago, Billington (1,2) in Sydney drew attention to an epidemic of chronic gastric ulcer which began in the early 1940's in N.S.W. and Queensland, affecting young women. The finding was apparently unique to Australia and the epidemiological data indicated an environmental cause; its nature was not known. The data indicated that the environmental influence, once it affected young women appeared to continue to exert an influence throughout that woman's life.

Soon after, Douglas and Johnston (3) in Townsville drew attention to the high frequency of women with chronic tension headaches taking large quantities of A.P.C. (aspirin phenacetin and caffeine preparations) powders among their patients with gastric ulcer. They speculated whether the aspirin in the A.P.C. was relevant and whether this had any bearing on the recently described epidemic.

In 1962, having noted a high rate of A.P.C. taking in my patients with bleeding gastric ulcer, I began a prospective study of patients admitted to the Royal Newcastle Hospital with acute perforated peptic ulcer - a readily defined emergency. In 118 patients seen over 4 years it was shown that there was a larger number of young women with gastric ulcer (4). Also, there was, in sufferers from perforated gastric ulcer, a significant association with chronic aspirin use (most often in the form of A.P.C. powders) when compared to duodenal ulcer sufferers (4).

About this time, a study of some 600 admission to the Royal Newcastle Hospital over 21 years by Dr.Hennessy showed a dramatic increase in the number of females with gastric ulcer during this period with a static rate for admissions of perforated duodenal ulcer in males.

About this time, Gillies and Skyring examined 50 admissions for chronic gastric ulcer and compared them with patients with duodenal ulcer, gastric cancer and with controls (5). They showed that of patients with gastric ulcer, more took aspirin regularly, in a larger dose and for a longer time in comparison with controls (5). In my unit we observed an association of bleeding gastric ulcer and chronic aspirin use. Subsequently, examination of the aspirin use of 295 patients admitted with peptic ulcer 1960-1966 revealed an excess of gastric ulcers in women aged 30-59 years (6). There was a statistically significant association with the consumption of large quantities of A.P.C. powders (6).

In the last few years, interest in the phenomena has been shown overseas. In the Boston Collaborative Drug Study (reported by Levy) comprising 25,000 admissions to hospital in Boston, there was evidence of a significant association of regular aspirin taking and uncomplicated gastric ulcer (7). More recently, Cameron published evidence of an association of regular aspirin use and gastric ulcer in outpatients compared to controls (8). A study of U.S.Veterans admitting with bleeding ulcer by Lee and Dagradi has extended the association by showing an increased prevalence of aspirin usage in those with multiple compared to single gastric ulcers (9).

In recent studies we have investigated all the cases of proven gastric ulcers in the community of 330,000 in Newcastle (Australia) over one year (10,11). Some of the relevant findings are these:
- (a) the incidence in females (144) is the same as that in males (146), a finding not found elsewhere in the world,
- (b) again a significant association of gastric ulcer in young women with aspirin taken regularly daily, generally in the form of A.P.C. powders,
- (c) 12 % of the patients were also taking other ulcerogenic drugs such as indomethacin or phenylbutazone.

It is suggested that the mode of action of these non-steroidal anti-inflammatory agents in producing gastric mucosal lesions lies in their common property of inhibiting prostaglandin synthetase (12). Prostaglandin E_1 and E_2 both depress gastric secretion (13,14) and promote gastric mucosal vasodilatation (15). It appears that inhibition of the enzyme could promote gastric secretion and simultaneously produces vasoconstriction. It is suggested that ischaemic necrosis may then ensue, so producing gastric muscosal loss which may then lead to the formation of a chronic gastric ulcer.

Summary:

(a) In eastern Australia, there is an epidemic of gastric ulcer which began 30 years in young women who are now middle-aged, (b) This epidemic is associated with the regular use of aspirin, most often in the form of A.P.C. powders taken for non-medical reasons, (c) Data from three separate studies in America confirms a statistically significant association of regular aspirin use and chronic gastric ulcer, and (d) An explanation is provided by the effect of aspirin in inhibiting prostaglandin synthetase.

REFERENCES

(1) B.P.BILLINGTON, Gastric Ulcer: Age, Sex and a Curious Retrogression, Aust.Ann.Med. 9, 111-121 (1960).

(2) B.P.BILLINGTON, The Australian Gastric Ulcer Change: Further Observations, Med.J.Aust. 2, 19-20 (1960).

(3) R.A.DOUGLAS and E.D.JOHNSON, Aspirin and Chronic Gastric Ulcer, Med.J.Aust. 2, 893-897 (1960).

(4) J.M.DUGGAN, The Relationship between Perforated Defitic Ulcer and Aspirin Ingestion, Med.J.Aust. 2, 659-662 (1965).

(5) M.GILLES and A.SKYRING, Gastric Ulcer, Duodenal Ulcer and Gastric Carcinoma: A Case-Control Study of Certain Social and Environmental Factors, Med.J.Aust. 2, 1132-1136 (1968).

(6) B.L.CHAPMAN and J.M.DUGGAN, Aspirin and Uncomplicated Peptic Ulcer, Gut 10, 443-450 (1969).

(7) M.LEVY, Aspirin Use in Patients with Major Upper Gastro-intestinal Bleeding and Peptic Ulcer Disease, New Engl.J. Med. 290, 1158-1162 (1974),

(8) A.J.CAMERON, Aspirin and Gastric Ulcers, Mayo Clin.Proc. 50, 565-570 (1975).

(9) E.R.LEE and A.E.DAGRADI, Haemorrhagic Erosive Gastritis, A Clinical Study, Am.J.Gastroent. 63, 201-208 (1975).

(10) D.PIPER and J.M.DUGGAN, Unpublished Studies.

(11) J.M.DUGGAN, Aspirin in Chronic Gastric Ulcer: An Australian Experience, Gut, in press (1976).

(12) R.FLOWER, R.GRYGLEWSKI, K.HERBACYZNSKA-CEDRO and J.R.VANE, Effects of Anti-Inflammatory Drugs on Prostaglandin Biosynthesis, Nature (New Biol.) 238, 104-196 (1972).

(13) A.ROBERT, J.E.NEZAMIS and J.P.PHILLIPS, Effect of Prostaglandin E_1 on Gastric Secretion and Ulcer Formation in the Rat, Gastroenterology, 55, 481-487 (1968).

(14) Y.H.LEE, W.D.CHENG, R.G.BIANCHI, K.MOLLISON and J.HANSEN, Effects of Oral Administration of PGE_2 on Gastric Secretion and Experimental Peptic Ulcerations, Prostaglandins 3, 29-45 (1973).

(15) D.F.HORRIBIN, M.S.MANKU, R.KARMALI, B.S.NASSAR and P.A.DAVIES, Aspirin, Indomethacin, Catecholamine and Prostaglandin Interactions on Rat Arterioles and Rabbit Hearts, Nature, Lond. 250, 425-426 (1974).

INTERACTIONS OF ANTI-RHEUMATIC DRUGS

by P.M.Brooks[1], M.A. Bell[2], D.I. Mason[3] and W.W.Buchanan[2]
1) Department of Medicine, University of Tasmania Medical School, G.P.O. Box 252C, Hobart, Tasmania, Australia 7001.
2) Centre for Rheumatic Diseases, Glasgow, Scotland
3) Royal Infirmary, Glasgow, Scotland

INTRODUCTION

The management of joint pain caused by osteoarthritis, rheumatoid arthritis or related disorders provides doctors throughout the world with a constant therapeutic challenge. The exact cause of these diseases remains obscure and treatment is still largely symptomatic. A whole host of therapeutic agents are now available to treat these diseases, and as all these agents produce side-effects rationalisation of drug therapy in this disease is of prime importance. In this paper we present data on in-patient and out-patient prescribing patterns at the Centre for Rheumatic Diseases in Glasgow, and in other hospitals in the West of Scotland. This will be contrasted with the prescribing patterns in North America kindly provided to us by the Boston Collaborative Drug Surveillance Programme (B.C.D.S.P.). Finally, some data will be presented on the interactions of these agents in patients with rheumatoid arthritis.

There are a number of reasons why a doctor chooses a particular therapeutic agent in preference to another agent having a similar effect and there is no doubt that advertising in the form of medical papers or drug promotional literature is important in trying to change a doctors prescribing habits. However, it is true to say that to change prescribing habits is difficult and our basic habits will depend on the principals of therapeutics that we have learnt at medical school and in our early professional life. To this end it is important that the teaching of therapeutics be patient oriented rather than drug oriented as it still is in many teaching institutions today.

Lee and coworkers ([1]) looked at the prescribing habits of General Practitioners in the Glasgow area in their treatment of patients with rheumatoid arthritis. They found that only one third of the patients received salicylates as the initial treatment and 15 % had not received them at all prior to consultation with the specialist unit. Of the patients who had been treated with salicylates nearly a third had to discontinue the treatment because of side effects and in nearly all these patients, the reason for discontinuation was dyspepsia. Phenylbutazone was the initial treatment of 10 % of patients, and corticosteroids were used initially in 5 % (though over a third of the patients had been given corticosteroids before referral to the specialist centre). The most common form of aspirin prescribed was soluble aspirin (in 80 % of the patients) and enteric coated aspirin was used in the remainders. Over 10 % of patients had been given two types of aspirin at the same time. It was after these initial

Figure 1.

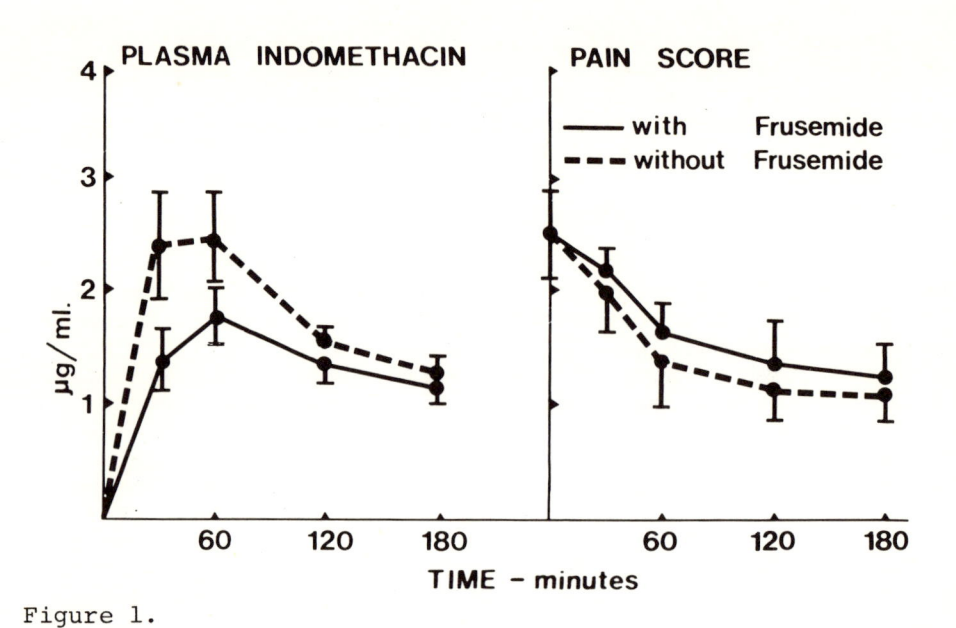

Indomethacin-Frusemide interaction. Plasma indomethacin and pain score profiles.

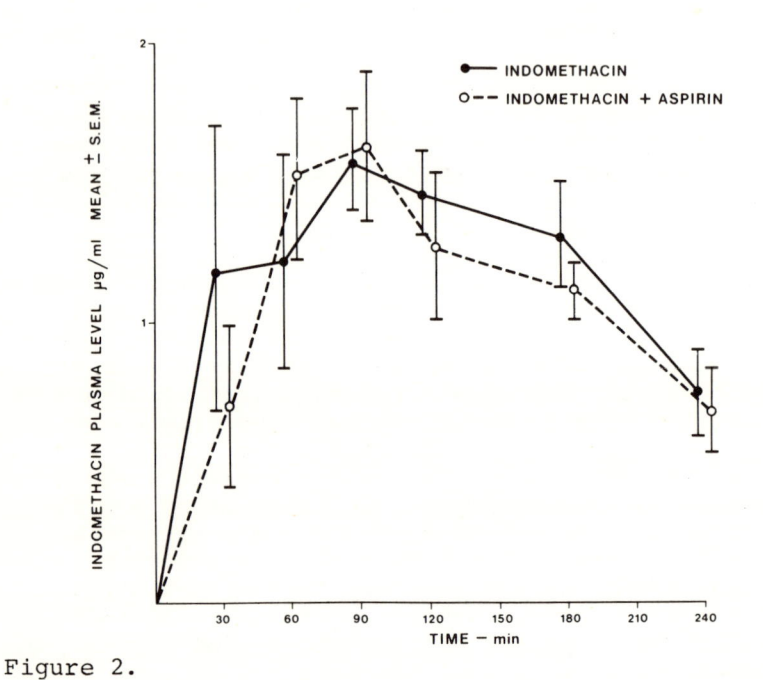

Figure 2.

Indomethacin-Aspirin interaction. Plasma indomethacin profiles.

studies that it was decided to investigate prescribing habits in a Specialist Centre, and in some other general hospitals in Scotland and America.

The drug treatment sheets of 153 consecutive admissions with rheumatoid arthritis to the Centre for Rheumatic Diseases, Glasgow, in 1973 (Group A), and 36 patients admitted to two other Glasgow teaching hospitals with rheumatoid arthritis (Group B), were studied. The prescribing sheets used were those described by Crooks and coworkers ([2]). These allow one to determine the number of drugs taken as a function of time. Drug prescribing data was then obtained in 164 patients with rheumatoid arthritis admitted to six American hospitals (Group C). This data was provided by the Boston Collaborative Drug Surveillance Programme, and was collected by trained observers who noted the exact time that drugs were commenced or discontinued allowing an analysis to be carried out to determine the maximum number of drugs taken concurrently. The average age of the patients in Group A was 54 years. The average duration of hospitalisation was 32 days, and 35 % were male. Comparable figures for Group B were 62 years, 23 days and 40 %; and for Group C were 52 years, 21 days and 60 %. The slightly higher incidence of males in the American data was due to the fact that several of the hospitals participating in the programme were Veteran's administration hospitals.

The average number of drugs consumed regularly, prior to admission by patients in these three groups is shown in Table 1.

TABLE 1

Number of drugs taken on admission - mean ± standard deviation.

Group	A	B	C
All Drugs	2.3 ± 1.6	2.4 ± 1.4	3.8 ± 2.3
N.S.A.I.D.	1.1 ± 0.6	0.7 ± 0.7	0.8 ± 0.5

Where the American patient consumed more drugs prior to hospitalisation than did their Scottish counterparts, the prevalence of non-steroidal anti-inflammatory drugs, (N.S.A.I.D.), used in America was less than in the Group A Scottish patients although similar to the Group B patients. The maximum number of drugs consumed at any one time, and the average number of N.S.A.I.D. and analgesic agents given to these patients is shown in Table 2.

Some 30 % of patients in both of the Scottish studies received two or more N.S.A.I.D. concurrently as compared with only 10 % of the patients in the American studies. One of the major factors in this difference may be that the American prescribers had only four N.S.A.I.D. to choose from, whereas Scottish prescibers had more than ten of these agents. There is at present little evidence to suggest that combinations are better than the use

TABLE 2

Drugs Taken While In Hospital (mean ± SD).

Group	A	B	C
Maximum No. of Drugs	4.6 ± 2.1	4.6 ± 1.9	5.9 ± 2.8
No. of N.S.A.I.D.	1.5 ± 0.7	1.0 ± 0.8	1.1 ± 0.3
No. of Analgesics	0.7 ± 0.9	0.9 ± 0.9	0.9 ± 0.7

TABLE 3

Indomethacin – Aspirin Study (mean ± S.E.M. : n = 14)

	Aspirin + Placebo	Indomethacin + Placebo	Indomethacin + Aspirin
Pain score	2.5 ± 0.2	2.4 ± 0.2	2.7 ± 0.2
Articular Index	17.7 ± 2.8	19.1 ± 2.5	17.2 ± 2.5
Patient Assessment	3.3 ± 0.3	3.1 ± 0.2	3.1 ± 0.2
Grip Stength (mm Hg.):			
Right hand	116 ± 8	110 ± 12	112 ± 7
Left hand	115 ± 9	106 ± 9	106 ± 8

of single N.S.A.I.D. for the treatment of rheumatoid arthritis and we feel that these drugs should be used alone rather than in combination. A third of the patients in the Glasgow hospitals were given aspirin in a daily dose of less than 3 g per day, whereas only 10 % of the patients in America received doses of less than 3 g per day. Boardman and Hart (3) showed that a dose of 3 g or more of salicylates per day were required to demonstrate anti-inflammatory effect in patients with rheumatoid arthritis. Though the authors demonstrated this effect by measurement of digital joint circumference (which is considered a less accurate measure of inflammation (4,5)) and it is considered that in treating rheumatoid arthritis doses of salicylate in excess of 3 g per day should be used. It was interesting to note that N.S.A.I.D. were used more frequently in the Specialist Centre, in Scotland, than the less specific analgesic agents. To be effective in relieving the pain of rheumatoid arthirits, a drug must have anti-inflammatory as well as analgesic effects (6). In the other Glasgow hospitals and in the American hospitals N.S.A.I.D. and analgesic agents were apparently used with equal frequency.

Regional differences in drug prescribing have been noted both in general (7) and also in the rheumatic diseases (8). The results of these in-patient surveys are interesting, because they reflect differences in drug usage, either because of availability of drugs, or perhaps as a particular preference of a certain drug by a physician or unit. Drug interactions described on the medisc (9), were noted in about 20 % of each group studied, though of 40 interactions occurring in the in-patients at the Centre of Rheumatic Diseases, only five had clinical effects of an interaction. Much has been recently written about the need for some form of quality control of medical care (10,11) and these studies suggest that the prescribing of drugs in rheumatoid arthritis may be a suitable subject for medical audit.

There is another aspect of drug therapy which has received little attention at present and that is the morbidity and mortality due to drugs. Jick and coworkers (12) have shown that adverse drug reactions occur in 5 % of all drug exposures and these adverse reactions lead to death in a small percentage of patients (13). Recently Lee and coworkers (14) noted that adverse affects of drugs accounted for 6 % of admissions to a rheumatic disease hospital in Scotland. Also anti-rheumatic drugs account for approximately 50 % of deaths reported to the Committee on Safety of Medicines in Britain (15). In a recent study of the cause of death in patients with rheumatoid arthritis in Glasgow (16), drugs did not seem to be implicated in the cause of death of 80 %, but were probably implicated in 6 % of cases. Almost certainly drugs were implicated in 13 % of 82 deaths where details on the cause of death were available.

"A desire to take medicine", as Sir William Osler (17) aptly said "is perhaps, the greatest feature which distinguishes man from other animals". The great demand by patients for pills, and potions is matched only by the "furor therapeuticus" of many physicians to treat every symptom with a new drug (18). In the Oslerian era, diagnosis was of prime importance to the physician, whereas today treatment whould be of equal importance. However, if we are honest with ourselves and examine how we deliver medicines to our patients, we have to conclude that perhaps we do not exercise the same care and attention that we pay to the diagnostic process. Although, lip service is paid to the idea that each dose of a drug is a new experiment (19) in practice the physician must be constantly reminded if the concept is not to loose its' significance.

Drug interactions make up only a small proportion of the reported adverse reactions, but may account for a disproportionate number of deaths. In study of nearly 10,000 patients 6.9 % of 3,600 reactions were found due to drug interactions (20). Shapiro and coworkers (13) have observed that one third of deaths attributed to drug treatment were a result of a drug interaction. Drug interactions occur frequently in patients with rheumatoid arthritis, because of the large number of drugs these patients take. What we have tried to do is to look at the clinical relevance of some of these interactions rather than looking specifically at changes in plasma drug concentrations.

Van Arman and his colleagues (21) have clearly demonstrated antagonism between aspirin and a variety of non-steroidal and steroidal anti-inflammatory drugs in their effects on inflammation in the rat adjuvant induced arthritis. An antagonism has also been shown between indomethacin and certain analgesic drugs (devoid of anti-inflammatory actions). It was not found necessary for the drug to be given together for this antagonism to occur and the antagonism produced was suprisingly long lasting. In carrageenen-induced foot swelling in the rat the affects of aspirin and indomethacin have not been found to be additive as might be expected (22,23). Pharmacokinetic interactions have been demonstrated between two new anti-inflammatory agents Naproxen (24) and Fenoprofen (25); the plasma concentrations of both drugs being significantly reduced when taken with aspirin.

There has been considerable controversy regarding the interaction of aspirin and indomethacin. Champion and coworkers (26) and Lindquist (27) have not found any significnat differences in indomethacin plasma levels when administered with aspirin. Recently Moller (using a more sensitive indomethacin assay) has shown a reduction in total serum indomethacin level but an increase in the free indomethacin level (28). The critical question in all these interaction studies is does a change in the plasma level of an anti-rheumatic drug make any difference to the pain relief of a patient with rheumatoid arthrirts? The assessment of a response to a drug is always difficult but using the subjective indices of pain relief, it is easier to measure responses of patients with rheumatoid arthritis to drugs than in some other conditions.

Deodhar and colleagues (4) have shown that of the methods available for measuring the clinical response to an anti-inflammatory drug therapy in rheumatoid arthritis, the most reliable were the patients' assessment of pain in their joints and the articular index of joint tenderness using the method of Ritchie and coworkers (29). Recently a new method for assessing the therapeutic potential of anti-inflammatory drugs using subjective criteria only has been described (30). Using this method the patient charts his degree of pain daily on a pain chart and assesses the treament at the end of a two week period. Significant differences can be shown between placebo and steroids but no difference can be shown between the non-steroidal anti-inflammatory drugs.

Probenecid has been shown to increase indomethacin blood concentrations by reducing renal tubular secretion. We have recently shown (32) that significant increases in plasma indomethacin levels after concurrent administration of indomethacin and probenecid are associated with a significant decrease in the duration of morning stiffness and in the articular index of patients with sera-positive rheumatoid arthritis. The clinical improvement in patients was seen whether the indomethacin was taken orally or by suppository. The interesting feature observed was that there was an extremely low incidence of side-effects of indomethacin; discontinuation of the drug was required only in five of sixty patients (despite the high plasma levels of indomethacin). It is conceivable that probenecid might counteract the side-effects of indomethacin and further studies are being undertaken to determine whether this might be the case.

Patients with rheumatoid arthritis are often given diuretics in an effort to reduce ankle swelling. To this end we investigated the interaction of frusemide and indomethacin (33). Pain profiles and plasma indomethacin profiles were repeated in eight patients with rheumatoid arthritis after the ingestion of 50 mg of indomethacin with and without 40 mg of frusemide. The results are shown in Figure 1 and indicate a significant decrease in plasma indomethacin concentrations without a significant change in the clinical state of the patient.

Indomethacin and aspirin are commonly prescribed together and we investigated this interaction in the clinic. Twenty patients with sero-positive rheumatoid arthritis (according to the American Rheumatism Association criteria) were included in the study and were given, in turn, two week treatment periods of (1) aspirin and indomethacin placebo, (2) aspirin placebo and indomethacin and (3) aspirin and indomethacin. The treatments were administered in a random order as determined by a latin square matrix. The results of the study (summarised in Table 3) show that there is no significant difference between the three treatment groups (34). Indomethacin plasma profiles were studied before and after a week's treament of soluble aspirin in a dose of 4 g per day in six normal volunteers. The profiles are shown in Figure 2 and again no significant difference is shown between the plasma profiles. Side-effects were experienced in 20 treatments (33 %) and the side effect necessitated cessation of therapy in 6 patients. Ten (50 %) of the adverse effects occurred when the patient was taking aspirin and indomethacin together. The numbers involved are small but most side-effects occur in patients taking both medications together and we feel that these non-steroidal anti-inflammatory agents should be used singly rather than in combination.

Recently, it has been shown that enzyme inducing agents can also alter pain levels particularly in rheumatoid patients maintained on corticosteroids (35). The addition of phenobarbitone to a corticosteroid regime in 9 patients with rheumatoid arthritis significantly increased their articular index of joint tenderness, pain score and morning stiffness. In a parallel study the tritiated prednisolone half life was observed to be singificantly reduced in these patients. For this reason a change in disease activity following introduction of additional therapy should be considered to have a iatrogenic component related to changes in established drug metabolism, rather than be dismissed casually as a naturally occuring relapse or remission.

In conclusion, we have demonstrated some of the problems in the prescribing of anti-rheumatic drugs and have shown some of the ways that these drugs can interact together. The clinical relevance of the interaction is the all important thing and we can only echo the words of Dollery (36) that "these problems can only be solved at the bedside and clinical pharmacology will not prosper if its practitioners spend too much time with the gas chromatography and too little time with the patients".

ACKNOWLEDGEMENTS

The authors wish to thank Dr.H.Jick and Dr.D.H.Lawson for data from the B.C.D.S.P. P.M. B. was in receipt of an A.H.Robins Co. Research Fellowship in Clinical Rheumatology.

REFERENCES

(1) P.LEE, S.J.AHOLA, D.M.GRENNAN, P.M.BROOKS and W.W.BUCHANAN, Observations on Drug Prescribing in Rheumatoid Arthritis, Brit.Med.J. 1, 424-426 (1974).

(2) J.CROOKS, C.G.CLARK, H.B.CAIE and W.B.MAWSON, Prescribing and Administration of Drugs in Hospital, Lancet 1, 373-378 (1965).

(3) P.L.BOARDMAN and F.D.HART, Clinical Measurement of Anti-Inflammatory Affects of Salicylates in Rheumatoid Arthritis Brit.Med.J. 4, 264-268 (1967).

(4) S.D.DEODHAR, W.C.DICK, R.HODGKINSON and W.W.BUCHANAN, Measurement of Clinical Response to Anti-Inflammatory Drug Therapy in Rheumatoid Arthritis, Quart.J.Med. 42, 387-401 (1973).

(5) J.WEBB, W.W.DOWNIE, W.C.DICK and P.LEE, Evaluation of Digital Joint Circumference Measurements in Rheumatoid Arthritis, Scand.J.Rheumatol. 2, 127-131 (1973).

(6) P.LEE, M.WATSON, J.WEBB, J.ANDERSON and W.W.BUCHANAN, Therapeutic Effectiveness of Paracetamol in Rheumatoid Arthritis, Int.J.Clin.Pharmac. 11, 68-75 (1975).

(7) D.DUNLOP and R.S.INCH, Variations in Pharmaceutical and Medical Practice in Europe, Brit.Med.J. 3, 749-752 (1972).

(8) Y.MIZUSHIMA, Method of Treatment of Rheumatoid Arthritis in Different Countries. Acta Rheum.Scand. 12, 210-218 (1966).

(9) B.WHITING, A.GOLDBERG and P.WALDIE, The Drug Disc System for Drug Interaction, Lancet 1, 1037-1038 (1973).

(10) H.W.K.ACHESON, Medical Audit and General Practice, Lancet 1, 511-513 (1975).

(11) P.SANAZARO, Medical Audit Experience in the U.S.A., Brit.Med.J. 1, 271-274 (1974).

(12) H.JICK, O.S.MIETTINEN, S.SHAPIRO, G.P.LEWIS, V.SISKIND and D.SLONE, Comprehensive Drug Surveillance, J.Am.Med.Assn. 219, 1455-1460 (1970).

(13) S.SHAPIRO, D.SLONE, G.P.LEWIS and H.JICK, Fatal Drug Reactions among Medical Inpatients, J.Am.Med.Assn. 216, 467-472 (1971).

(14) P.LEE, S.MCCUSKER, A.ALLISON and G.NUKI, Adverse Reactions in Patients with Rheumatic Diseases, Ann.Rheum.Dis. 32, 565-573 (1973).

(15) R.H.GIRDWOOD, Death after taking Medicaments, Brit.Med.J. 1, 501-504 (1974).

(16) P.B.BROOKS, W.H.STEPHENS, M.E.B.STEPHENS and W.W.BUCHANAN, How Safe are Anti-Rheumatic Drugs? A Study of Possible Iatrogenic Deaths in Patients with Rheumatoid Arthritis, Health Bulletin 33, 108-111 (1975).

(17) W.OSLER, Science 17, 170 (1891).

(18) D.DUNLOP, Abuse of Drugs by the Public, and by Doctors, Brit.Med.Bull 26, 236-240 (1970).

(19) J.P.QUILLIAM and P.TURNER, Clinical Pharmacology - Its Role and its Integration into the Education of the Medical Student, Lancet 2, 1081-1083 (1967).

(20) I.T.BORDA, D.SLONE and H.JICK, Assessment of Adverse Reactions within a Drug Surveillance Programme, J.Am.Med.Assn. 205, 645-647 (1968).

(21) C.G.VAN ARMAN, G.W.NUSS and E.A.RISLEY, Interactions from Aspirin, Indomethacin and other Drugs in Adjuvant - Induced Arthritis in the Rat, J.Pharmacol.Exp.Ther. 187, 400-414 (1973).

(22) Z.E.MIELLENS, H.P.DROBECK, J.Jr.ROZITIS and S.J.Jr.SANSONE, Interactions of Aspirin with Non-Steroidal Anti-Inflammatory Drugs in Rats, J.Pharm.Pharmacol. 20, 567-569 (1968).

(23) K.F.SWINGLE, T.J.GRANT, L.W.JAQUES and D.C.KVAM, Interactions of Anti-Inflammatory Drugs on Carageenan - Induced foot Oedema of the Rat, J.Pharmacol.Exp.Ther. 172, 423-425 (1970).

(24) E.J.SEGRE, M.CHAPLIN, E.FORCHIELLI, R.RUNKEL and H.SEVELIUS, Naproxen-Aspirin Interactions in Man, Clin.Pharmacol.Ther. 15, 374-379 (1974).

(25) A.RUBIN, B.E.RODDA, P.WARRICK, C.M.Jr.GRUBER and A.S.RIDOLFO, Interactions in of Aspirin with Non-Steroidal Anti-Inflammatory Drugs in Man, Arthr.Rheum. 16, 635-645 (1973).

(26) D.G.CHAMPION, H.E.PAULUS, E.MONGAN, R.OKUN, C.M.PEARSON and E.SARKISSON, The Effect of Aspirin on Serum Indomethacin, Clin.Pharm.Ther. 13, 239-244 (1972).

(27) B.LINDQUIST, K.M.JENSEN, H.JOHANSSAN and T.HANSEN, The Effect of Concurrent Administration of Aspirin and Indomethacin on Serum Concentration, Clin.Pharmacol.Ther. 15, 247-252 (1974)

(28) P.W.MOLLER, Drug Interactions with Anti-Rheumatic Drugs, Current Therapeutics Feb.1976, 41-47 (1976).

(29) D.M.RITCHIE, J.A.BOYLE, J.M.MCINNES, M.K.JANSANI, T.G.DALAKOS P.GRIEVESON and W.W.BUCHANAN, Clinical Studies of an Articular Index for Assessment of Joint Tenderness in Patients with Rheumatoid Arthritis, Quart.J.Med. 37, 393-406 (1967).

(30) P.LEE, J.WEBB, J.A.ANDERSON and W.W.BUCHANAN, Method for Assessing Therapeutic Potential of Anti-Inflammatory Anti-Rheumatic Drugs in Rheumatoid Arthritis, Brit.J.Med. 2, 685-688 (1973).

(31) M.D.SKIETH, P.A.SIMKIN and L.A.HEALEY, The Renal Excretion of Indomethacin and its Inhibition by Probenecid, Clin. Pharmacol.Ther. 9, 89-93 (1968).

(32) P.M.BROOKS, M.A.BELL, R.D.STURROCK, J.P.FAMAEY and W.C.DICK, The Clinical Significance of Indomethacin-Probenecid Interaction, Brit.J.Clin.Pharmacol. 1, 287 (1974).

(33) P.M.BROOKS, M.A.BELL, P.LEE, P.J.ROONEY and W.C.DICK, The Effect of Frusemide of Indomethacin Plasma Levels, Brit.J. Clin.Pharmacol. 1, 485-489 (1974).

(34) P.M.BROOKS, J.J.WALKER, M.A.BELL, W.W.BUCHANAN and A.R. RHYMER, Indomethacin-Aspirin Interaction - A Clinical Appraisal, Brit.Med.J. $\underline{3}$, 69-71 (1975).

(35) P.M.BROOKS, W.W.BUCHANAN, M.GROVE and W.W.DOWNIE, The Effects of Enzyme Induction on the Metabolism of Prednisolone - A Clinical and Laboratory Study, Ann.Rheum.Dis., In press (1976).

(36) C.T.DOLLERY, Editorial: Pharmaco-Kinetics - Master or Servant? Eur.J.Clin.Pharmacol. $\underline{6}$, 1-2 (1973).

SALICYLATE TOXICOLOGY

by E.G. McQueen
Department of Pharmacology, University of Otago Medical School,
Dunedin, New Zealand

I suppose that nothing has ever come closer to being a panacea in our Anglo-Saxon Society than aspirin. People of British descendance have come to take it for every conceivable ailment, real or imaginary. They take it for feverishness, sore throat, rheumatism, all of which are perfectly reasonable and respectable reasons for taking it. But additonally they have come to take it for "nerves", which is silly, or dyspepsia, which could be disastrous. Sometimes they take it for no explicable reason at all other than that all the girls at work take it, and their mothers and aunts take it, and they would have looked a bit peculiar if they didn't take it too! This idea has been assiduously fostered by the advertising industry.

As a consequence there has never been a drug with such ubiquitous distribution. Every home has it; its in every typist's drawer; every teacher has it; it's in _every_ factory sick bay. It would be remarkable really if every now and then it did not do somebody quite a lot of no good!

And the first and most obvious way in which it may do harm is through poisoning, both accidental poisoning of infants and deliberate self-poisoning.

1. Metabolism and Distribution in Relation to Toxicity:

Aspirin, and occasionally methyl salicylate, are the only salicylates now clinically encountered in Australasia. They are rapidly absorbed, at least in part from the stomach, and hydrolysed in the blood to free salicylic acid. However, in the case of aspirin at therapeutic concentrations it would appear that a proportion of a dose of aspirin circulates as such for at least an hour or so and one wonders what significance this may have. Lester and coworkers ([1]) found that at 30 minutes after a dose of 0.625 g of aspirin, 27 % of the total salicylate was present as aspirin. Levy and Leonards ([2]) quote data from their own earlier work indicating that at rapid rates of absorption (such as would no doubt apply with a massive overdose) that a substantial fraction of the dose would be circulating initially as acetylsalicylic acid. The contribution of acetylsalicylic acid to analgesic effect in excess of that accountable as salicylate ([3]) suggests that aspirin as a separate species exerts pharmacological, and therefore no doubt toxicological, effects in its own right.

Salicylate circulates bound to a variable degree to serum albumin. The extent of the latter in our own experiments using 4 % human albumin would vary, within the range of clinical toxicity, from 70 % at 480 µg/ml to 44 % at 1240 µg/ml of salicylate respectively. Aspirin is bound to a much smaller extent. Deficiency of serum albumin as a result of prior inanition or of renal or hepatic disease will of course modify the interpretation of serum levels of salicylate; being of graver significance in the hypoalbuminaemic patient because less will be retained in the plasma compartment. Probably not enough is sequestered ordinarily in plasma in the bound form for deficient binding to acutally increase the toxicity of a given dose. If at blood levels of say 1000 µg/ml salicylate the binding were lowered from 50 % to 25 %, the amount liberated by the reduction in binding would only be about 0.75 g (i.e 2 ½ tablets). Another possibility that may cause diminished binding and influence toxicity would be the displacement of salicylate from protein binding which could occur with increasing concentration of acidic metabolites. As mentioned earlier the relatively low level of binding of aspirin would enable drug circulating as aspirin to leave the vascular compartment relatively freely.

Distribution of salicylate depends upon pH of body fluids. Volume of distribution (Vd) is about 0.15 l/kg in the normal subject, that is to say about equivalent to the extracellular fluid volume. However in acidic poisoned patients, Vd is probably much higher. At pH 7.4, salicylate with a pKa of 3.0 is 99.996 % ionized, i.e. 0.004 % unionized. At pH 6.9 0.0125 % is unionized, i.e. three times as much than at 7.4. Therefore, presumably the rate of passage of salicylate across membranes, including the blood brain barrier, would be three times as fast as pH 6.9 as at pH 7.4. Acetylsalicylic acid with a slightly higher pKa (3.5) would be expected to penetrate slightly faster than salicylate. In a series of infants with salicylate poisoning investigated by Buchanan and Rabinowitz ([4]), a strong correlation was observed between the serum to cerebro-spinal fluid (CSF) salicylate ratio and the actual serum bicarbonate level. Salicylate enters the CSF relatively slowly, but ultimately the CSF concentration will equilibrate with the free (unbound) salicylate in the plasma. In this ([4]) series, the mean CSF level was a little higher than the serum free, presumably because the fall of salicylate in the serum had been more rapid than that in the CSF. The extent of entry into the cerebral tissue will depend upon the ratio of intracellular pH to plasma pH. Since the intestinal cellular fluid (ICF) is usually more acidic than plasma, the degree of intracellular ionisation will be less than that outside. Hence the retention of salicylate within cells will be discouraged. However, at very low plasma pH there may be an equivalent degree of ionisation in each compartment which could promote a substantial movement into the ICF, particularly into cerebral tissue. Proudfoot and Brown ([5]) observed a severe impairment of consciousness in acidaemic adult patients poisoned with salicylate than in non acidaemic patients with similar salicylic levels.

Elimination of salicylate at toxic levels is preponderantly by renal excretion of unmetabolized salicylate, with a half life of 15-30 hours ([6]). At these extreme levels the capacity of salicylurate and glucuronide systems will have been grossly exceeded and their activity quite probably suppressed in salicylate toxicity. Elimination does not follow 1st order kinetics until the body content is down to about 400 mg aspirin equivalent (6). Renal elimination will be greatly modified by the fraction of drug in the tubule which is in the ionized state. A tenfold increase in clearance of unconjugated salicylate occurs with a 2 unit rise in urine pH (7). The presence of osmotically active molecules in the filtrate will also enhance elimination by trapping water and reducing the gradient for salicylate reabsorption.

2. Toxic Manifestations:

The major toxic effects of salicylates may be grouped as follows:

(1) Local gastrointestinal
(2) Central nervous system
(3) Metabolic
(4) Haematological

(1) **Local Gastrointestinal Effects**: These include vomiting, the vomitus often containing blood, and severe substernal and epigastric pain. Vomiting contributes to dehydration.

(2) **Central Nervous System (CNS) Effects** also induce vomiting by stimulation of the CNS. Tinnitus and deafness are initial subjective complaints going on to delirium, convulsions and ultimately coma. The respiratory centre is both directly stimulated by salicylates and rendered more sensitive to pH changes resulting from the metabolic disturbances (6). Extreme hyperpnoea persists throughout the clinical courses of the poisoning episode. Direct stimulation of the respiratory centre causes an initial alkalosis which is often demonstrable in adults, but seldom in children, in whom a metabolic acidosis predominates in the early stages. The hyperpnoea also contributes to the dehydration.

Initial agitation may be accompanied by a toxic psychosis with paranoid and hallucinatory behaviour and asterixis. Delirium supervenes with possibly epileptiform seizures and finally coma. Proudfoot and Brown (5) reported that provided no sedative drugs had been ingested simultaneously, the impairment of consciousness invariably was associated with acidaemia.

(3) **Metabolic Effects**: These include a large increase in metabolic rate associated with increased heat production which induces fever or even hyperpyrexia in infants. This results from the action of salicylate in uncoupling mitochondrial oxidative phosphorylation. There is a marked increase in O_2 consumption, CO_2 production, tachycardia and increased cardiac output (6).

Defective energy production resulting from the uncoupling is additionally accompanied by derangement of carbohydrate meta-

bolism with either hyperglycaemia or hypoglycaemia (6). There
may possibly be a reduction in glucose due in the CNS to increased cerebral glycolysis compensating for the uncoupled oxidative phosphorylation in the absence of systemic hypoglycaemia
(8). Salicylate-poisoned animals treated with glucose-free
fluids died rapidly with hypoglycaemic convulsions (7). Deranged
carbohydrate metabolism and dehydration are potent stimuli to
ketosis and metabolic acidosis. Excretion of ketone acids,
Kreb's cycle acids and amino acids depletes the bicarbonate
buffer system and actual acidaemia may supervene. Acidaemia is
almost invariably a manifestation of metabolic acidosis and
results from interference by salicylate with metabolic pathways
leading to the accumulation of organic anions, but the precise
mechanism is not understood. If renal function is already impaired, or if the patient is an analgesic abuser, and already
has substantial salicylate tissue levels, development of acidosis may be accelerated. Circulatory failure and consequent
acute renal insufficiency will of course enhance the rate of
development of acidaemia even in the absence of prior pathology.
Acidaemic patients are usually equally as hypocapnic as the
non-acidaemic ones and hence CO_2 retention is not an important
pathogenetic factor. Where respiration is markedly depressed by
other drugs it may contribute to the acidosis. Urinary reaction
is likely to be acidic even in the presence of alkalosis, hypokalaemia no doubt contributing to this (9). Hypokalaemia seems
to be invariable and K^+ administration is necessary from the
outset if K^+ deficit is to be avoided. K^+ loss will be markedly
enhanced if acetazolamide is used.

Water deficits from vomiting and hyperpnoea plus initial urinary
losses may lead to hyperosmolarity. Untreated, hypotension with
oliguria and acute renal failure are likely.

The net disturbance of hydrogen ion concentration in salicylism
is the product of the two simultaneous challenges to acid-base
homeostasis. CNS-mediated hyperpnoea and respiratory alkalosis;
deranged carbohydrate metabolism with ketosis, oligaemia and
impaired renal function combine to produce metabolic acidosis.
In adults the patient generally presents with a respiratory alkalosis and acidosis may develop as the clinical course progresses. In young children, due to their greater susceptibility
to ketosis, the state is one of metabolic acidosis from the
time of presentation. Buchanan and Rabinowitz in an extensive
study of ten infants with severe salicylism, were able to des-
cribe what might have been a mixed respiratory alkalosis/meta-
bolic acidosis in one; all the others had a metabolic acidosis
on presentation (4).

If CNS depressant drugs are ingested at the same time hyperpnoea
may be markedly suppressed with enhanced rate of development of
acidosis and a severe impairment of cerebral function. Such combinations as dextropropoxyphene and aspirin are peculiarly
dangerous (10). If phenacetin is involved as well, as in aspirin-
phenacetin-codeine poisoning which I have seen a few times, the

signs on presentation may be dominated by the methaemoglobinaemia resulting from the phenacetin with intense cyanosis. Since the codeine suppressed respiration the picture appears quite different from the usual salicylate one. The association of aspirin and paracetamol either in a compound preparation or as a chemical combination (Benorylate) would also seem to be particularly hazardous.

(4) <u>Haematological Effects</u>: The coagulation defect associated with salicylism is compounded of a series of induced deficits including increased capillary fragility, impaired platelet aggregation, decreased Factor VII levels and hypoprothrombinaemia. Occasionally there may be life-threatening complications such as an intracerebral bleed or severe gastrointestinal haemorrhage.

3. Management:

Emergency: fluids++; emesis. Hospital: gastric lavage. Leave cholestyramine (15 g) plus sodium sulphate (15-20 g) in stomach. Estimate plasma salicylate. At 35 mg% = poisoning; 80 mg% = prognosis bad. Rate of increase also influences severity. Estimate also pCO_2, bicarbonate, blood and urine pH. Forced diuresis with alkali enhances urinary elimination and minimizes brain levels.

<u>Adults</u>: i.v. Mannitol (20%) 75 ml   <u>Children</u>: 1 ml/kg Mannitol
         i.v. Frusemide 40 mg   (as appropriate)

Follow with i.v. fluids in rotation

(a) 500 ml 5 % dextrose + 50 mEq $NaHCO_3$)
(b) 500 ml 5 % dextrose + 25 mEq.KCl   )   at rate of
(c) 500 ml physiological saline   )   30 ml/kg/hr

Control convulsions. Watch for and treat hyperpyrexia. Respiratory tract toilet. Prophylactic antibiotic. Vitamin K 10 mg i.m. In very severe cases haemodialysis. In infants exchange transfusion may be of value.

The most dangerous complication of salicylism is acidaemia. The essential feature of management is the supply of fluid, base and glucose.

Initiation of measures for elimination must await control of acidaemia, if present, or reassurance that it is not. Hence priorities may be rather different in adults from those in children. Since a respiratory alkalosis is usual in adults it is appropriate to give particular attention to alkalinizing the urine at an early stage, whereas in infants the major priority is to secure control of the acidaemia.

I believe an initial dose of mannitol together with i.v. frusemide is appropriate to initiate diuresis concomitantly with the commencement of the i.v. fluid regime. However, great care must be taken to avoid pulmonary oedema and further i.v. fluids must be contingent upon establishment of a satisfactory urine flow. Central venous pressure should be monitored if there is any dubiety. Pulmonary oedema may be a sequel of salicylate poisoning even in the absence of fluid overload (11).

The use of acetazolamide has generated a good deal of controversy. I think there is sufficient evidence of increased risk in children to consider that it should be contraindicated. On the other hand, in adults where the plasma base reserve is well established and the pH relatively high it will enhance elimination rate substantially. I would be happy to use it in adults if the urinary pH is persistently low but blood pH relatively high. It might be helpful to a degree in compensating for systemic alkalosis. Serum K^+ should be monitored closely and additional potassium given as indicated.

4. Prophylaxis of Acute Poisoning:

This requires that a responsible attitude towards the care of medicine be adopted both privately by parents in the home and generally by the community. The responsibility of parents embraces fully adequate custody of medicines as well as a responsible attitude towards the domestic use of aspirin. A substantial proportion of any series of cases of salicylism results from the administration of aspirin to infants in doses which parents and often medical attendants fail to appreciate as being potentially dangerous (12). Measures taken by the community via its regulatory authorities can significantly modify the incidence of poisoning in infants. Since the introduction of measures to encourage the use of safety containers in the United States, there has been a substantial reduction in the number of death from salicylate poisoning (Table 1).

In New Zealand, by contrast, the number of children admitted to hospital for salicylate poisoning has remained about the same over the last three years for which figures are available (Table 2).

Regulations gazetted recently in New Zealand require that aspirin and a considerable range of other medicines be packaged individually in foil wrapped packaging. The results, hopefully, will become evident shortly. Requirement for use of safety closures for therapeutic substances in liquid form is currently under consideration; methyl salicylate being the product carrying the greatest hazard in this context.

The overall frequency of salicylate poisoning is decreasing both in Australia (13) and in New Zealand (see Table 2) because of diminishing popularity amongst adults for self-poisoning.

5. Chronic Poisoning and Adverse Drug Reactions:

(1) _Gastrointestinal Reaction_: These are quantitatively more important than those involving all the other systems together, but enough has already been said at this Symposium on this aspect.

TABLE 1

Deaths of Children under 5 from Accidental Poisoning, U.S.A. 1968-1973.

(Data from National Clearinghouse for Poisons Control Centre)

Year	All drugs	Analgesic/antipyretic	Salicylates
1968	150	73	61
1969	137	68	58
1970	124	58	48
1971	141	57	44
1972	142	64	46
1973	102*	37*	26*

* Safety packaging regulations' introduced latter part of 1972.

TABLE 2

"Salicylates and Congeners" N.Z.

(Data from NZ Department of Health)

Cases Presenting to Public Hospitals (Deaths in Brackets)

Year/Age	0 - 5	5 - 15	15 - 25	25+	Total
1972	59	21	167	82	329 (1)
1973	69	23	198	71	342 (0)
1974	58	13	135	57	263 (3)

(2) **Aspirin and the Kidney:** Analgesic nephropathy has been a disaster in Australia (14) and a source of greater or lesser morbidity in various parts of the world (15); the causes of the differences in frequency not being entirely apparent. Published statistics would not suggest a high incidence in New Zealand but I have seen about twenty cases. With few exceptions the patients have conformed to the pattern described by Kincaid-Smith (14) and also many have misused a variety of drugs other than analgesics. They have mostly been women but about a quarter have been men. Characteristics they shared are an incapacity to tolerate discomfort either physical or mental, and a requirement for instant succour which places intolerable demands upon their medical attendants. The condition carries a high mortality rate, which in my patients has been fairly evenly distributed between renal failure and suicide, which even extensive psychiatric assistance has failed to preclude.

The lesions characterising analgesic nephropathy have been more than adequately described (14) but their is much debate on the role of aspirin. There is no doubt that persons who take compound analgesic tablets in excess can suffer with analgesic nephropathy, but patients may be on substantial doses of salicylate for prolonged periods without any obvious effect on renal function (16). Furthermore, phenacetin-containing compound analgesics not including aspirin consistently produce impairment of renal function (17). Papillary necrosis can regularly be produced in rats by aspirin particularly if reinforced by fluid restriction, but the unipapillary rat kidney may not be an appropriate model. All the twenty cases of analgesic nephropathy I have had under my care have at some time taken large amounts of phenacetin; the majority of these showing collateral evidence of phenacetin moridity on presentation in the form of methaemoglobinaemia, often with evidence of haemolysis.

However, there is also consistent evidence of a degree of renal damage produced by aspirin in high dosage as manifested by an increased rate of cell excretion (18) and of minor functional impairment, particularly that involving tubular function (19, 20). N-acetyl-B-D-glucosaminidase excretion, which is considered a sensitive test of renal damage, is also increased. The results (20) suggest that salicylate treatment does cause renal tubular damage but that this damage results in only minimal impairment of function.

(3) **Hepatotoxic Effects:** These are an aspect of salicylate toxicity which is infrequently seen but may be a source of diagnostic confusion and therapeutic difficulty in patients with lupus erythematosis (21). Lupoid hepatitis may alternatively be suggested for it may be difficult to distinguish the histological appearance between presumed salicylate-induced hepatotoxicity and lupoid hepatitis. Hepatotoxic manifestations may appear in association with other rheumatic disorders in which the use of aspirin may be employed such as acute rheumatic fever (22) and juvenile rheumatoid arthritis (23).

(4) **Allergy to Aspirin:** The first description of an allergy to aspirin appeared over 74 years ago, only three years after it had been synthesized. Its frequency is said to be as high as 4% in asthmatics (24). Manifestations include urticaria, purpura, vasomotor rhinitis, asthma, nausea, vomiting, diarrhoea, angioedema and even sudden death (6). Symptoms may come on immediately after taking aspirin or be delayed. Amongst subjects with overt allergys bronchoconstriction is the most prominent manifestation in asthmatics and urticaria the most common in subjects with rhinitis. However, it is not peculiar to atopic subjects. Settipane and coworkers (24) found that aspirin intolerance occurred in 0.9% of subjects without any other evidence of allergy, and was manifested equally by urticaria and bronchospasm. The frequency increases with increasing age. Allergic subjects with aspirin-sensitivity often have nasal polyps. Removal of polyps does not relieve the aspirin sensitivity and may enhance the severity of bronchoconstrictor response. The group of subjects characteristically affected are middle aged women with nasal polyps or chronic sinusitis having a known history of atropy and eosinophilia. It has been postulated that since aspirin-sensitive patients may develop reactions after other analgesic-anti-inflammatory drugs such as indomethacin, antipyrine, aminopyrine and phenylbutazone, the aspirin-reaction may not be mediated through an antibody-antigen reaction. A possible suggested mechanism is via differential block in the synthesis of the bronchodilator prostaglandin PGE_2; this possibly being inhibited more than the synthesis of the bronchoconstrictor prostaglandin, $PGF_{2\alpha}$.

However sensitivity of this kind does not apparently occur with salicylate (25), whereas aspirin is more potent as an inhibitor of prostaglandin synthesis *in vitro*, sodium salicylate is effective at a similar concentration in human subjects (26), as might be expected from the rapid hydrolysis of aspirin *in vivo*. Nevertheless the duration of the inhibitory action of aspirin on platelet prostaglandin biosynthesis is much longer (2-3 days) than that of sodium salicylate (6 hours) (27).

The ability of aspirin to acetylate proteins (28) would seem to endow it with a potent capacity to act as a hapten and consequently as an allergen. Studies of possible causative immune mechanisms have given evidence of the antigenicity of aspiryl chloride in rats and guinea pigs although similar studies in human aspirin-sensitive subjects have given negative results (29).

(5) **Maternal and Foetal Effects:** A most disquieting picture of the attrition exacted by the headache powder habit in New South Wales was brought to light by an investigation of their maternal and foetal effects carried out at Crown Street Women's Hospital in Sydney (30, 31). The regular takers of analgesic powders had an increased requency of anaemia, antepartum and post-partum haemorrhage, prolonged gestation and complicated deliveries when compared with non-takers. Amongst the infants of the analgesic-

takers there was a significant reduction in birth-weight and the stillbirth and perinatal death rates were significantly higher. The obstetric complications were suggested to have related to the prostaglandin inhibitory effects of the salicylates as was the increased frequency of post-partum haemorrhage, together with diminished platelet aggregation. The latter could have contributed also to the more frequent anaemia in the analgesic group who might additionally have been expected to have gastric erosions. In this study it was found the brand most often used then contained phenacetin as well as aspirin.

Perhaps the most disconcerting aspect of the study was that of the Australian-born women attending the antenatal clinic, 6.6% were found to be taking analgesics containing salicylates regularly.

(6) Teratogenicity: Although there have been occasional reports of this association they are infrequent and without any consistent pattern of deformity so that it seems unlikely to be important.

(7) Other Aspects:

(a) Fluid retention: Cardiac failure is occasionally precipitated in patients with acute rheumatic fever as a result of fluid retention. The occurrence of pulmonary oedema in patients poisoned with salicylates has already been mentioned.

(b) Ototoxicity: Delayed recovery of hearing or even permanent deafness has been reported after toxic or rarely therapeutic doses of salicylates.

(c) Thyroid: Salicylates lower protein bound iodine and displace thyroxine from plasma pre-albumin. They may confuse interpretation of tests but do not seem to produce clinical effects.

(d) Uric acid: At doses of 1-2 g daily, salicylates inhibit renal excretion of uric acid and cause hyperuricaemia. At higher dosage effects are uricosuric but there may be interaction with other drugs such as sulphinpyrazole the effects of which are inhibited.

(e) Interaction of drugs: Other than the above, interaction of clinical significance involve particularly anticoagulants and oral andidiabetic drugs, by displacement from binding and/or by enhancement of pharmacological effect.

REFERENCES

(1) D.LESTER, C.LOLLI and L.A.GREENBERG, The Fate of Acetylsalicylic Acid, J.Pharmacol.Exp.Ther. 87, 329-342 (1946).

(2) G.LEVY and J.R.LEONARDS, Absorption, Metabolism and Excretion of Salicylates. In: The Salicylates (Ed. M.J.H. Smith and the late P.K.Smith, Interscience, New York, 1966) pp. 5-48.

(3) H.O.J. COLLIER, A Pharmacological Analysis of Aspirin. Advances Pharmacol.Chemotherap. 7, 333-405 (1969).

(4) N.BUCHANAN and L.RABINOWITZ, Infantile Salicylism - A Reappraisal. J.Pediat. 84, 391-395 (1974).

(5) A.R.PROUDFOOT and S.S.BROWN, Acidemia and Salicylate Poisoning in Adults. Brit.med.J. 2, 547-550 (1969).

(6) M.J.H.SMITH and P.K.SMITH, The Salicylates. A Critical Bibliographic Review. Interscience, New York, London, Sydney (1966).

(7) E.W.REIMOLD, H.G.WORTHEN and T.P.REILLY, Salicylate Poisoning, Am.J.Dis.Child. 125, 668-674 (1973).

(8) J.H.THURSTON, P.G.POLLOCK, S.K.WARREN and E.M.JONES, Reduced Brain Glucose with Normal Plasma Glucose in Salicylate Poisoning, J.Clin.Invest. 49, 2139-2145 (1970).

(9) A.W.PIERCE, Salicylate Intoxication, Postgrad.Med. 48, 243-249 (1970).

(10) R.D.WARREN, D.S.MEYERS, B.A.PAPE and J.F.MAHER, Fatal Overdose of Propoxyphene Napsylate and Aspirin, J.Am.Med.Assn. 230, 259-260 (1974).

(11) G.HRNICEK, J.SKELTON and W.C.MILLER, Pulmonary Edema and Salicylate Intoxication, J.Am.Med.Assn. 230, 866-867 (1974).

(12) F.T.SHANNON, Aspirin Medication in Infancy and Childhood, N.Z.Med.J. 64, 571-573 (1965),

(13) D.J.MCCLEAVE and J.HAVILL, A Review of Salicylate Poisoning, Anaesth.Intens.Care 4, 340-344 (1974).

(14) P.KINCAID-SMITH, Analgesic Nephropathy. A Common Form of Renal Disease in Australia, Med.J.Aust. 2, 1131-1135 (1969).

(15) J.A.ABEL, Analgesic Nephropathy - A Review of the Literature, 1967-1970, Clin.Pharmacol.Therap. 12, 583-598 (1971).

(16) N.Z.RHEUMATISM ASSOCIATION, Aspirin and the Kidney, Brit. med.J. 1, 593-596 (1974).

(17) U.C.DUBACH, P.S.LEVY, B.ROSNER, H.R.BAUMELER, A.MULLER, A. PEIER and T.EHRENSPERGER, Relation between Regular Intake of Phenacetin-Containing Analgesics and Laboratory Evidence for Uro-Renal Disorders in a Working Female Population in Switzerland, Lancet 1, 539-543 (1975).

(18) J.T.SCOTT, A.M.DENMAN and J.DORLING, Renal Irritation Caused by Salicylates, Lancet 1, 344-348 (1963).

(19) T.W.STEELE, A.Z.GYORY and K.D.G.EDWARDS, Renal Function in Analgesic Nephropathy. Brit.med.J. 2, 213-216 (1969).

(20) H.C.BURRY, P.A.DIEPPE, F.B.BRESNIHAN and C.BROWN, Salicylates and Renal Function in Rheumatoid Arthritis, Brit.med.J. 1, 613-615 (1976).

(21) W.E.SEAMAN, K.C.ISHAK and P.H.PLOTZ, Aspirin-induced Hepatotoxicity in Patients with Systemic Lupus Erythematosus, Ann.Int.Med. 80, 1-18 (1974).

(22) T.IANCU, Serum Transaminases and Salicylate Therapy, Brit.med.J. 2, 167 (1972).

(23) G.M.KOPPES and F.C.ARNETT, Salicylate Hepatotoxicity, Postgrad.Med. 56, 193-195 (1974).

(24) G.A.SETTIPANE, F.H.CHAFEE and D.E.KLEIN, Aspirin Intolerance II: A Prospective Study in an Atopic and Normal Population. J.Allergy and Clin.Immunol. 53, 200-204 (1974).

(25) M.SAMTER and R.F.BEERS, Intolerance to Aspirin, Ann.Intern.Med. 68, 975-983 (1968).

(26) M.HAMBERG, Inhibition of Prostaglandin Synthesis in Man, Biochem.Biophys.Res.Commun. 49, 720-726 (1972).

(27) J.J.KOCSIS, J.HERNANDOVICH, M.J.SILVER, J.B.SMITH and C.INGERMAN, Duration of Inhibition of Platelet Prostaglandin Formation and Aggregation by Ingested Aspirin or Indomethacin. Prostaglandins 3, 141-145 (1973).

(28) D.HAWKINS, R.N.PINCKARD and R.S.FARR, Acetylation of Human Serum Albumin by Acetylsalicylic Acid, Science 160, 780-781 (1968).

(29) B.GIRALDO, M.N.BLUMENTHAL and W.W.SPINK, Aspirin Intolerance and Asthma, Ann.Intern.Med. 71, 479-496 (1969).

(30) E.COLLINS and G.TURNER, Maternal Effects of Regular Salicylate Ingestion in Pregnancy. Lancet 2, 335-337 (1975).

(31) G.TURNER and E.COLLINS, Foetal Effects of Regular Salicylate Ingestion in Pregnancy, Lancet 2, 338-339 (1975).

TERNARY COPPER (II) COMPLEXES CONTAINING SALICYLATE AND NITROGENOUS CHELATES SUCH AS HISTAMINE.

by W.R.Walker and R.Reeves
Department of Chemistry, The University of Newcastle, Shortland, N.S.W., 2308, Australia

Although the effect of salicylic acid on rheumatic fever has been known for 100 years, the relationship between its anti-inflammatory action and its ability to chelate has been noted for only 25 years (1). It is also of interest that 10 years have elapsed since Schubert (2) first suggested that salicylates delivered copper to the body cells. Indeed, this role of copper in enhancing the potency of non-steroid anti-inflammatory drugs is only now receiving attention (3,4,5). A brief survey of the structures of copper (II) aspirinate and salicylate complexes is therefore relevant.

The structure of copper (II) aspirinate has been shown by Manojlovic-Muir (6) to possess a polymeric structure containing dimeric units similar to that of copper (II) acetate. It is important to note that whereas this copper (II) aspirinate complex is insoluble, the adducts (ternary complexes) $(Cu(aspirinate)_2 DMF)_2$ and $(Cu(aspirinate)_2 DMSO)_2$ (DMF = dimethylformamide, DMSO = dimethylsulphoxide) are soluble in dioxane (7).

Complexes of bivalent copper and salicylates have been known for many years. Hanic and Michalov (8) have prepared $Cu(Hsal)_2 \cdot nH_2O$ (n = 4,2,1 and 0) and have shown that both the dihydrate and tetrahydrate are not chelated. The coordination is square-planar with two carboxylate oxygen donors at 1.84 Å and two oxygens of water at 1.92 Å. Magnetically different crystal modifications have been prepared and described by Inoue and coworkers (9).

The nature of these complexes is also pH dependent and may differ when in solution. For example, at pH 7 the compound $CuSal \cdot H_2O$ results. Its structure has not been reported. It is an unusual yellow-brown in colour and is believed to be a polymeric chelate. At pH 8 the green compound $Na_2(Cu(sal)_2) \cdot 2H_2O$ is formed. It is the most soluble complex and is also thought the contain the salicylate dianion as a bidentate chelate (10).

The stability constants of some metal salicylate complexes have been reported by Perrin (11). Values for Cu^{2+} are log K_1 = 10.60 and log K_2 = 7.85 (other M^{2+} ions are ≃ 7 and ≃ 5). For Fe^{3+}, the values log K_1 = 16.35 and log K_2 = 11.90, were reported. (The incidence of severe anaemia amongst salicylate users is not surprising when these results are considered).

Copper (II) complexes containing mixed-ligands continue to attract attention especially in biological systems. Huber and coworkers (12) have reported stability constants for mixed species such as the tenary complex Cu-histamine-pyrocatechol and noted that the imidazole-containing complexes are unusually stable. They stated that "the first coordinated ligand has an influence on the kind of the second to be coordinated i.e. the copper-histamine 1:1 complex shows discriminating properties". Other studies have been made by Tanaka (13) and by Martin and Prados (14). The crystal structures of copper (II) with mixed imidazole and glycine peptide ligands have been reported by Bell et al. (15).

As part of an investigation of the interaction of Cu^{2+} with biologically important molecules, imidazole (16,17) histamine and histidine (18) have received particular attention. Because of experience with binuclear hydroxy-bridged copper (II) complexes such as shown in structure (I)

$$(\text{chelate } Cu \overset{OH}{\underset{OH}{\diagup\!\!\!\diagdown}} Cu \text{ chelate})^{2+}$$

(I)

where chelate = Ha (18,20), phen and bipy (19), it was decided to attempt replacing the bridging hydroxyl ligands by the two oxygen donors of the salicylate dianion. Because our previous studies (20) showed that species such as structure (I) may be important in the physiological activity of histamine it seemed that the mixed complex Ha-Cu-sal, may be of particular interest. This paper describes the mixed salicylate complexes that were subsequently prepared.

Experimental:

(1) Preparation of mono-complexes - Cu chelate Cl_2

$CuHaCl_2$ was prepared as described earlier (18), and Cu phen Cl_2 and Cu bipy Cl_2 were prepared by adding one mole of the chelate dissolved in aqueous ethanol to one mole of copper (II) chloride dissolved in water. The crystalline complexes were dried over P_2O_5 (vac) and the analyses may be seen in Table 1.

TABLE 1

Elementary Analyses (%)

Compound	Colour		Cu	C	H	N
$CuHaCl_2$	blue-green	Found	25.2	24.9	3.9	17.7
		Calc.	25.9	24.5	3.7	17.5
Cu phen Cl_2	pale green	Found	20.1	46.0	2.7	8.4
		Calc.	20.2	45.8	2.6	8.9
Cu bipy Cl_2	pale green	Found	-	41.4	3.0	9.0
		Calc.	-	41.3	2.8	9.6

TABLE 2

Elementary Analyses (%)

Compound	Colour		Cu	C	H	N
(Ha-Cu-sal)°	green	Found	19.2	43.9	4.3	13.0
		Calc.	19.3	43.8	4.6	12.8
(phen-Cu-sal)°	green	Found	15.9	57.1	3.5	6.8
		Calc.	16.0	57.4	3.5	7.0
(bipy-Cu-sal)°	pale green	Found	17.0	52.0	4.1	6.9
		Calc.	16.2	52.1	4.1	7.2

TABLE 3

Diffuse Reflectance Specta of Solid Complexes

Compound	λ_{max} (nm)	Compound	λ_{max} (nm)
$(Cu(phen)_2)(ClO_4)_2 \cdot H_2O$	780	$(Cu\text{-histamine})_2)(ClO_4)_2$	555
$(Cu(pipy)_2)(ClO_4)_2$	720	$Na_2(Cu(sal)_2) \cdot 2H_2O$	615

	λ_{max} Calculated		λ_{max} Found
(Cu Ha sal)°	585		600
(Cu phen sal)°	697		600
(Cu pipy sal)°	667		585

(2) **Preparation of ternary complexes - (chelate-Cu-Sal)°**

The ternary complexes were prepared by the following procedure. The mono-complex was dissolved in water, one mole of aqueous NaOH was added followed by one mole of sodium salicylate. The extremely insoluble products were filtered off, washed and dried. Analyses are shown in Table 2.

(3) *In vivo* experiments, with white mice, were performed as was described earlier (20). A stock solution of Cu Ha Cl$_2$ was prepared in 0.9 % saline and the pH was adjusted to 7 with NaOH. A solution of sodium salicylate in aqueous DMSO was added to this solution in order to make a stock solution of Cu:Ha:sal (1:1:1) of 0.3 M concentration in 70 % DMSO. That this stock solution

TABLE 4

Computation (by method of Reed and Muench described in Ref.20) of LD_{50} for $(Cu\ Ha\ sal)^o$ in 70 % DMSO; LD_{50} is dose to kill 50 % of mice via anaphylactic shock within 10 min.

Dose mixed complex 70% DMSO μmoles/g mouse	Mortality Ratio	Died	Survived	Total Dead	Total Survived	Ratio	%
3.60	10/10	10	0	40	0	40/40	100
2.21	14/18	14	4	30	4	30/34	88.2
1.55	8/12	8	4	16	8	16/24	66.7
1.13	7/22	7	15	8	23	8/31	25.8
0.77	1/10	1	9	1	32	1/33	3.0
0.59	0/4	0	4	0	36	0/36	0

LD_{50} = 1.32 μmoles/g mouse.

TABLE 5

Effect of sodium salicylate on histamine response in mice [†]

Dose of sod.sal. (μmoles/g mouse)	Mortality ratio
4	1/18
2	3/14

[†] The sodium salicylate was injected prior to injection of the LD_{50} dose of histamine (20).

was identical to that obtained by dissolving the pure compound $(CuHasal)^o$ in DMSO was proven by UV - visible spectrometry (λ_{max} = 600 and ε = 100 in both cases).

The effect of sodium salicylate on the pharmacological action of histamine in mice was also investigated. The results of all of these in vivo studies appear in Tables 4 and 5.

Results and Discussion:

These three mixed-salicylate copper (II) complexes were prepared from the corresponding mono-complexes Cu chelate Cl_2 (chelate = Ha, phen and bipy) by the addition of sodium hydroxide followed by sodium salicylate. As stated in the introduction this was planned, because it had been shown (18,20) that the following equilibria are involved:

$$2CuChelate^{2+} \rightleftharpoons (Cu(Chelate)_2)^{2+} + Cu^{2+} \quad \ldots\ldots (1)$$

$$(Cu(Chelate)_2)^{2+} + Cu^{2+} \underset{2H_2O}{\rightleftharpoons} (chelate\ Cu \overset{OH}{\underset{OH}{\diagup\diagdown}} Cu\ chelate)^{2+} + 2H^+. (2)$$

This paper suggests that the claim by Huber and coworkers (12) for the discriminating qualities of the Cu^{2+} - histamine 1:1 complex may be better explained by invoking binuclear-hydroxy-bridged species of structure (i). The formation of these mixed-salicylate complexes could then be explained by the reaction:

$$(chelate\ Cu \overset{OH}{\underset{OH}{\diagup\diagdown}} Cu\ chelate)^{2+} + 2Hsal^- \rightarrow 2(chelate\ Cu\ sal)^o +$$

$$2H_2O \ldots\ldots (3)$$

Evidence for the rearrangement (1) may be found in the molecular conductivities of aqueous (2×10^{-3}M) solutions of Cu chelate Cl_2. At room temperature they are 230, 215 and 210 ω respectively for Ha, phen and bipy. Spectral data also support these equilibria (1) and (2). When it is recalled that such hydroxy-bridged species were prepared (19) by the following reaction:

$$2Cu\ chelate^{2+} + 2OH^- \rightarrow (chelate\ Cu \overset{OH}{\underset{OH}{\diagup\diagdown}} Cu\ chelate)^{2+} \quad \ldots\ldots (4)$$

the proposed mechanism (reaction 3) for the formation of the salicylate-mixed complexes seems probable.

The possibility of these ternary salicylate complexes being of the type $(Cu(chelate)_2)(Cu(sal)_2)$ has not been overlooked. In Table 3 are listed spectral data for complexes of the type $(Cu(chelate)_2)^{2+}$ taken from Sone and coworkers (21) along with λ_{max} of the solid ternary complexes and that of $Na_2(Cu(sal)_2)$. $2H_2O$. These permit the calculation of a theoretical λ_{max} for the dimeric species $(Cu(chelate)_2)(Cu(sal)_2)$.

The differences in the calculated and observed values for λ_{max} fo the ternary complexes (Cu sal chelate)^o preclude the possibility of them being of the type $(Cu(chelate)_2)(Cu(sal)_2)$. For this reason and because of their method of preparation, these highly insoluble complexes may possess structure (II).

This formulation is in keeping with earlier studies. The insolubility of the 1:1:1 complex, (Cu(II) phen sal)^o, was observed by Condik and Martell (22). Further, the stability constant of (Cu(II) bipy sal)^o was determined by L'Heureux and Martell (23) to possess the value log K = 10.91 and that of (Cu Ha sal)^o was determined by Perrin (24) to be log K ~ 9.

TABLE 6

Clinical results obtained by Hangarter with a copper salicylate preparation * in the treatment of rheumatic fever, rheumatoid arthritis and sciatica (from Sorenson (28)).

	Symptom free	Improved	Slightly improved	Unchanged
Acute Rheumatic fever 78 patients	78 (100%)			
Rheumatoid arthritis 620 patients	403 (65 %)	143 (23 %)		74 (12 %)
Sciatica 120 patients without lumbar involvement	76	38	6	
160 patients with lumbar involvement	95	39	10	16 †
	171 (61 %)	77 (28 %)	16 (5.5 %)	16 (5.5 %)

† Twelve patients withdrew to undergo surgery.

* (a) <u>Intravenous injection</u>: each dose was a single ampoule containing 20 ml of an aqueous solution of sodium salicylate (2.0 g) and copper (2.5 mg).

  (b) <u>Intravenous infusion</u>: of 500 ml of isotonic saline containing 3 to 4 ampoules of above. (Treatment (b) adopted later than treatment (a)).

The ternary complex (Cu Ha sal)^0 may be relevant to the pharmacological actions of both histamine and the salicylates. If our hypothesis (20) is tenable, viz., that the physiological action of endogenous histamine involves copper and the following equilibria:

$$2Cu\ Ha^{2+} \rightleftharpoons (Cu\ (Ha)_2)^{2+} + Cu^{2+}$$

$$\updownarrow 2H_2O$$

$$(Ha - Cu \underset{OH}{\overset{OH}{\diagup\diagdown}} Cu - Ha)^{2+} + 2H^+ \quad \ldots (5)$$

then the function of salicylate may involve a similar antagonistic reaction to that caused by zinc. We have suggested (20) that the hydroxy-bridged species could not form in the presence of Zn^{2+} due to the formation of insoluble $Zn(OH)_2$ and histamine hydrochloride. Results of in vivo experiments involving $(Cu\ Ha\ sal)^0$ are shown in Table 4.

The above LD_{50} (1.32 μ moles/g mouse) is comparable to that for $Cu\ Ha\ Cl_2$ obtained earlier (20) and was 1.14 μ moles/g mouse. It is worth noting that the LD_{50} for histamine alone was 18.97 μ moles/g mouse (20). This unexpected result may be discussed in the light of equilibria (1) to (5) (where chelate = Ha). However, it must be appreciated that in vitro and in vivo reactions may not be comparable. For example, although $(Cu\ Ha\ sal)^0$ is extremely insoluble in water is is soluble in DMSO and this may be relevant to the solubilization of Cu(II) aspirinate by adduct formation (7,5). The similarity of the two LD_{50} values indicates that in both compounds, an equivalent amount of histamine per mole of copper is exerting its pharmacological effect.

The results of the in vivo experiments examining the effect of sodium salicylate on histamine alone appear in Table 5.

These data are consistent with previously reported observations of the antihistamine activity exerted by salicylates, including histamine induced injury to the capillary wall (25), histamine shock in the guinea pig (26) and vascular constriction in the isolated rabbit ear (27).

Significant evidence for the therapeutic value of copper salicylate in the treatment of arthritic disease is based on the work of Hangarter and relevant data are shown in Table 6.

SUMMARY

In the introduction the chemistry of some bivalent copper complexes of aspirin and salicylate is briefly reviewed and the biological importance of mixed-ligand complexes of bivalent copper is illustrated. The nature of hydroxy-bridged copper (II) complexes and their possible role in the physiological activity of histamine is also discussed.

Several neutral, insoluble copper (II) complexes of the type $(Chelate-Cu-salicylate)^0 \cdot x\ H_2O$ where chelate = 1,10-phenanthroline (phen), $x = 1$; 2,2'-bipyridyl(bipy), $x = 2$; histamine (Ha), $x = 1$; and where salicylate (sal) = dianion of salicylic acid have been prepared for the first time. They have been characterised by physico-chemical methods and the role of binuclear-hydroxy-bridged copper (II) complex ions in their formation is demonstrated. Such complexes may be relevant to the pharmacological action of histamine and of the salicylates, the copper complexes of which are potential anti-inflammatory drugs. Results of some in vivo experiments that have been carried out with mice are also presented.

The results of clinical trials carried out by Hangarter that show the therapeutic value of copper salicylate in rheumatic diseases, will also be discussed.

REFERENCES

(1) J.REID, R.D.WATSON, J.B.COCHRAN and D.H.SPROULL, Sodium ⨯-Resorcylate in Rheumatic Leyer, Brit.Med.J. 2, 321-326 (1951).

(2) J.SCHUBERT, Chelation in Medicine, Scient.Amer. 214 40-50 (1966).

(3) J.R.J.SORENSON, J.Med.Chem., in press (1976).

(4) K.D.RAINSFORD and M.W.WHITEHOUSE, Concerning the Merits of Copper Aspirin as a Potential Anti-Inflammatory Drug, J.Pharm.Pharmac. 28, 83-86 (1976).

(5) D.A.WILLIAMS, D.T.WALZ and W.O.FOYE, Synthesis and Biological Evaluation of Tetrakis (acetylsalicylate)-µ-dicopper(II) J.Pharm.Sciences 65, 126-128 (1976).

(6) L.MANOJLOVIC-MUIR, Copper (II) Acetylsalicylate, Chem.Comm. 1057-1058 (1967).

(7) K.S.BOSE and C.C.PATEL, Preparation and Infrared Spectra of Some Dimeric Copper (II) Arenecarboxylate Complexes, Indian J.Chem. 8, 840-842 (1970).

(8) F. VON HANIC and J.MICHALOV, Die Kristallstruktur von Kupfersalicylat-Tetrahydrat $Cu(OH.C_6H_4COO)_2.4H_2O$, Acta Cryst. 13, 299-302 (1960).

(9) M.INOUE, M.KISHITA and M.KUBO, Magnetically Different Crystal Modifications of Copper (II) Salicylate, Acta Cryst. 16, 699-700 (1963).

(10) G.A.POPOVICH et al., Electron Paramagnetic Resonance and Magnetic Susceptibility of Copper (II) Salicylates, Zh. Neorg.Khim. 14, 2710-2713 (1969).

(11) D.D.PERRIN, Stability of Metal Complexes with Salicylic Acid and related Substances, Nature Sept.13, 741-742 (1958).

(12) P.R.HUBER, R.GRIESSER, B.PRIJS and J.SIGEL, Ternary Complexes in Solution: The Stability Increasing Effect of the Imidazole Group on the Formation of Mixed Cu^{2+} Complexes, Eur.J.Biochem. 10, 238-242 (1969).

(13) M.TANAKA, The Mechanistic Consideration on the Formation Constants of Copper (II) Complexes, J.Inorg.Nucl.Chem. 36, 151-161 (1974).

(14) R.B.MARTIN and R.PRADOS, Some Factors Influencing Mixed Complex Formation, J.Inorg.Nucl.Chem. 36, 1665-1670 (1974).

(15) J.D.BELL, H.C.FREEMAN, A.M.WORD, R.DRIVER and W.R.WALKER, Crystal Structures of three Complexes of Copper (II) with Mixed Imidazole are Glycine Peptide Ligands, Chem.Comm. 1441-1443 (1969).

(16) M.E.BRIDSON and W.R.WALKER, Imidazole-Bridged Complexes of Copper (II), Aust.J.Chem. 23, 1973-1979 (1970).

(17) K.COLYVAS, R.P.COONEY and W.R.WALKER, Laser Raman and Infared Spectral Studies on Imidazolium Complexes of Bivalent Copper and Zinc, Aust.J.Chem. 26, 2059-2062 (1973).

(18) W.R.WALKER, YEUH-HO L.SHAW and C.W.NORMAN, Histidine and Histamine Complexes of Copper and Zinc, J.Coord.Chem. 3, 77-84 (1973).

(19) C.M.HARRIS, E.SINN, W.R.WALKER and P.R.WOOLIAMS, Nitrogen Chelate Complexes of Transition Metals, V. Binuclear Hydroxy-Bridge Copper (II) Complexes of 1,10-Phenantholine and 2,2'-Bipyridyl, Aust.J.Chem. 21, 631-640 (1968).

(20) W.R.WALKER, R.REEVES and D.J.KAY, The Role of Cu^{2+} and Zn^{2+} in the Physiological Activity of Histamine in Mice, Search 6, 134-135 (1975).

(21) K.SONE, S.UTSONO and T.OGURA, Absorption Spectra of Some Mixed Chelates of Copper (II), J.Inorg.Nucl.Chem. 31, 117-126 (1969).

(22) G.F.CONDIKE and A.E.MARTELL, Mixed Ligand Chelates of Copper (II), J.Inorg.Nucl.Chem. 31, 2455-2466 (1969).

(23) G.A.L'HEUREUX and A.E.MARTELL, Mixed Ligand Chelates of Copper (II), J.Inorg.Nucl.Chem. 28, 481-491 (1966).

(24) D.D.PERRIN, I.G.SAYCE and V.S.SHARMA, Mixed Liqand Complex Formation by Copper (II) Ion, J.Chem.Soc. (A) 1755-1759 (1967).

(25) G.I.M.SWYER, Anti-Histamine Effect of Sodium Salicylate and its Bearing Upon the Skin-Diffusing Activity of Hyaluronidase, Biochem.J. 42, 28-32 (1948).

(26) L.ZICHA and B.BERGULLA, Comparative Animal Experimental Studies on the Antagonistic Effect of Phenylbutazone and Oxyphenbutazone towards Histamine, Serotonin and Acetylcholine, Arzneim.-Forsch. (Drug Res.) 12, 474-477 (1962).

(27) R.DOMENJOZ, The Pharmacology of Phenylbutazone Analogues. In: Non-narcotic Drugs for the Relief of Pain and their Mechanism of Action, Ann.N.Y.Acad.Sci. 86, 263-291 (1960).

(28) J.R.J. SORENSON, personal communication.(1976).

ACKNOWLEDGEMENT

We thank the following companies for their generous financial assistance in aid for this conference:
- Bayer Pharmaceutical, Australia P/L
- Beecham Australia P/L
- The Boots Company of Australia P/L
- Nicholas Pty Ltd.
- Reckitts Pharmaceutical Ltd.